Alexander Arnfinn Olsen

Failure Mode and Effects Analysis

Guidance for the Marine and Offshore Industries

Alexander Arnfinn Olsen
Southampton, UK

ISBN 9781739171568

This Nørdik Press imprint is published by Nørdik Academia.

If disposing of this product, please recycle the paper.

Alexander Arnfinn Olsen

Failure Mode and Effects Analysis

Guidance for the Marine and Offshore Industries

Preface

The main Classification Societies (Class) often require their clients to develop and submit FMEA as part of the Classification requirements for select systems. For instance, FMEA are required for achieving many of the special or optional Classification notations such as the CDS, ACC, ACCU, R1, RQ, DPS-2, DPS-3, and ISQM offered by the *American Bureau of Shipping* (ABS) or other equivalents. This book provides guidance and insight into the development process for FMEA to comply with Classification rule requirements. The utilisation of this guidance will provide tangible benefits as the marine and offshore industry is able to realise the positive results of FMEA that are developed correctly and managed appropriately throughout the life cycle of a given system. Some of these benefits include:

- FMEA that meet the intended objectives and are a support to the classification process.
- Consistency in scope, depth and quality among comparable FMEA.
- Expedited FMEA review process; and
- Reduced failures, downtimes and incidents.

Terms of Use
The information presented herein is intended solely to assist the reader in the methodologies and/ or techniques discussed. The information herein does not and cannot replace the analysis and/ or advice of a qualified professional. It is the responsibility of the reader to perform their own assessment and obtain professional advice. Information contained herein is considered to be pertinent at the time of publication, but may be invalidated as a result of subsequent legislations, regulations, standards, methods, and/or more updated information and the reader assumes full responsibility for compliance.

Contents

1 Introduction 11
    Background 11
    Purpose of the FMEA..11
    Fmea Overview 12

2 Before the FMEA 15
    Preparing for the FMEA..15
    FMEA Standards 15
    Design Philosophy and FMEA...16
    FMEA Scope and Ground Rules 17
    Equipment Scope and Physical Boundaries 18
    Operational Boundaries (Global and Local) 22
    Failure Criteria and Types of Failure 22
    Depth of Analysis 22
    Criticality Ranking (FMECA)..23
    FMEA Naming Convention within this Book...23
    United States Coast Guard Supplemental Requirements for Qualitative Failure Analyses
     (QFA) 25
    FMEA Team 25
    Stakeholder Workshop Setting 26
    Third-Party FMEA Practitioner(s) 26
    Class Participation in the FMEA Workshop...26
    Team Preparation...27
    Ideal Timing to Conduct FMEA..27

3 Developing the FMEA 29
    Developing the FMEA 29
    Data Management 29
    FMEA Study...30
    Developing the Analytical Approach 30
    Functional Block Diagrams and Reliability Block Diagrams...............................34
    Identify Failure Modes..36
    Analyse Effects...40
    Identify Failure Detection Methods 41
    Identify Existing Risk Control Methods 42
    Criticality Ranking (for FMECA)..42
    Identify Corrective Actions 42

4 FMEA Report and Class Review of the FMEA...45
  FMEA Report..45
  FMEA Report Structure 45
  FMEA Internal Review Process 46
  Classification Review of the FMEA..46
  Pitfalls and Common Problems in Class Submitted FMEA..50
  FMEA and Supporting Documentation Submission 50
  Cause and Effects Matrix 51
  Operational Manuals (Operations, Inspection and Maintenance, and Emergency)............52

5 FMEA Verification Programme..53
  Purpose...53
  Scope of the FMEA Verification Programme...53
  Verification Programme Test Sheets...55
  Performing FMEA Verification Programme..55
  Results and Recommendations...56
  FMEA Verification Programme Report...56
  United States Coast Guard Design Verification Test Procedure..56

6 FMEA Lifecycle Management 59
  Best Practices for treating the FMEA as a Living Document 59
  Best Practices for treating the FMEA as an Operations Resource Document 60
  Best Practices for using the FMEA in Asset Lifecycle Management 60
  Changes to the Classification System and FMEA Revisions and Submissions....................62
  FMEA and Management of Change...62

7 System-Specific FMEA Requirements 65
  Guidance for System-Specific FMEA Requirements..65
  Automation (General Control, Safety-Related Functions of Computer-Based Systems,
    Wireless Data Communication, Integrated Automation Systems)....................................69
  Electronically Controlled Diesel Engines 77
  Remote Control Propulsion...81
  Gas Turbine 85
  Redundant Propulsion and Steering 89
  Single Pod Propulsion 93
  Dynamic Positioning Systems (DPS)..97
  Software Control System 107
  Jacking Systems 117
  Subsea Heavy Lifting..121
  Drilling System/Subsystems/Individual Equipment 125
  Integrated Drilling Plant...131
  Dual Fuel Diesel Engines (DFDE)...141
  Gas-Fuelled Engines 145
  Motion Compensation and Rope Tensioning Systems for Cranes.....................................151

Annex 155

Definitions, Abbreviations, and Acronyms

DEFINITIONS

Closed Bus. Closed buses often describe an operational configuration for a Dynamic Positioning System (DPS) where all or most sections and all or most switchboards are connected together, that is, the bus-tie breakers between switchboards are closed. The alternative to closed bus is open bus, sometimes called split bus or split ring. Closed buses are also called joined bus, tied bus, or closed ring.

Common Cause Failure. Failure of multiple components in a system as a result of a single cause.

Component Failure. Failure of individual hardware or component items. For purposes of this book, a component is a part that is easily replaceable on location. An FMEA addressing component failures is much more detailed than an FMEA that addresses failures at a functional level.

Computer-based System. A system of one or more programmable electronic devices, associated with software, peripherals, and interfaces. Microprocessors, Programmable Logic Controller (PLC), Distributed Control Systems (DCS), PC or server-based computation systems are examples of computer-based systems.

Detection Method. Means by which a failure is detected.

Enhanced System Notation (EHS). Enhanced System (EHS) notation, as a supplement to the Dynamic Positioning Systems (DPS)-series notation, may be assigned to a DPS-2 or DPS-3 vessel. The main objective of the enhanced system notation is to improve reliability, operability, and maintainability of the DP vessel. The enhanced system notations mainly emphasise the following properties of the DP system:

1. Robust design of power plants and thruster system.
2. Enhanced protection systems for generators and thrusters:
3. Failure detection and discrimination of failed components before a full or partial blackout situation occurs.
4. Quick automatic blackout recovery.
5. Enhanced position reference system and sensor system for increased availability and reliability.
6. Fire and flood protection for higher risk areas.

The notation includes provisions for standby start, closed bus operation and transferable generators. These are beneficial to the overall environment, operational flexibility, and system maintainability. Three separate enhanced notations are provided as follows:

- *Enhanced Power and Thruster System (EHS-P):* This notation covers the requirements for the power system and thrusters that are beyond those for the DPS-series notations.
- *Enhanced Control System: (EHS-C):* This notation covers the requirements on the DP control systems including control computers, position reference systems and sensors, which are beyond the minimum requirements for DPS-series notations.
- *Fire and Flood Protection System (EHS-F):* This notation covers the requirements for fire and flood protection considering the risk level of the areas. This is a supplement for a DPS-2 system. It is not necessary for a DPS-3 system since a DPS-3 system has higher requirements in this regard.

Equipment Under Control. Equipment that receives command from the control system.

Essential Service. Essential Services are those considered necessary for:

1. Continuous operation to maintain propulsion and steering (primary essential services)
2. Non-continuous operation to maintain propulsion and steering and a minimum level of safety for the vessel's navigation and systems including safety for dangerous cargoes to be carried (secondary essential services).
3. Emergency services.

Failure Cause. Defects in design, process, quality, or operation which have led to failure, or which initiate a process which leads to failure. For example, a circuit breaker contact was manufactured out of specification, or a pump was operated in a thermal environment beyond its design capability.

Failure Mode. Manner in which a component fails functionally. For example, a pump fails to provide pressure, or a circuit breaker fails to open.

Failure Modes and Effects Analysis (FMEA). A systematic process to identify potential failures to fulfil the intended function, to identify possible failure causes so the causes can be eliminated, and to locate the failure impacts so the impacts can be reduced.

Failure Modes, Effects and Criticality Analysis (FMECA). A variation of the FMEA that includes an explicit estimation of the severity of the consequences of the failure or a combination of the likelihood of the failure and severity of the consequences.

Fault Tolerance. Ability of a system to keep functioning to a predetermined level of satisfaction in the presence of faults or failures. Single fault tolerance refers to the ability of system to keep functioning to a predetermined level of satisfaction in the presence of a single fault or failure.

Fault Tolerant System. A system that is able to keep functioning to a predetermined level of satisfaction in the presence of faults or failures. A fault-tolerant system is designed from the ground up for reliability by building multiple independent critical components. In the event one component fails, another seamlessly takes over.

Firmware. Firmware is the combination of persistent memory and programme code, and data stored in it. The lowest level of firmware component failure that should be included in the FMEA is each easily replaceable component.

FMEA Testing. See "Verification Programme."

FMEA Trials. See "Verification Programme."

Functional Failure. Failures pertain to a particular function of the equipment not being performed or performed incorrectly. For example, for a system that needs to pump x gpm from point A to point B, typical functional failures would include failure of pumping capability, pumping at a rate below requirements, pumping at a rate exceeding requirements and pumping backwards.

The causes or failure mechanisms for these functional failures would include motor failure; loss of power; degraded pump or motor; under voltage to motor; overvoltage to motor; leaky non-return valve on discharge of pump. In order to perform a functional failure FMEA, the functions of the item under review must be defined. Note that the system/equipment under review may have more than one function.

Global Effects. The total effect an identified failure has on the operation, function or status of the installation or vessel and end effects on safety and the environment.

Global Operations. Global operations are the overall operations of the facility or vessel.

Hardware. In ISQM terms, hardware refers to the control system of physical electronic devices hardware or network such as:

HIL Testing. Hardware-in-the-loop (HIL) simulation tests control algorithms without using actual systems.

Integrated Automation. An Integrated Automation System (IAS) is a combination of computer-based systems with redundant architecture which are interconnected in order to allow communication between computer systems; between computer systems and monitoring, control, and vessel management systems; and to allow centralised access to information and/or command/control. For example, an integrated system may consist of systems capable of performing passage execution (e.g., steering, speed control, traffic surveillance, voyage planning); machinery management and control (e.g., power management, machinery monitoring, fuel oil/lubrication oil transfer); cargo operations (e.g., cargo monitoring, inert gas generation, loading/discharging); etc. Functions are integrated to reduce the need for hardware and software functions and to reduce interface requirements. For Integrated Automated Systems, the following design philosophies for Classification must be complied with to avoid hazardous situations:

1. All active devices have local/integrated control over detailed functions.
2. Control of the device/machine is limited to functions permitted by the local integrated control (limited instruction set). Machines look after their own internal safety.
3. All data/indications/alarms are available to the Integrated Automation System
4. Integrated automation systems cannot give commands to devices that will override the safety or produce dangerous outcomes.

If the philosophy is not complied with, the details of the means of controlling the individual devices and associated safety are to be submitted for consideration.

Integrated Software Quality Management (ISQM). ISQM is a risk-based software development and maintenance process built on internationally recognised standards. The ISQM process verifies the software installation on the facility, places emphasis on the verification of the integration of multiple software packages, and monitors for consistency when there are software updates or a change in hardware.

Local Effects. Impacts that an identified failure mode has on the operation or function of the item under consideration or adjacent systems.

Local Operations. Local operations are the operations of the system that is within the boundaries of the FMEA.

Operational Modes. High-level system(s) configurations for a specific set of operations. See "Local Operations" and "Global Operations."

Owner. Entity that owns the vessel or offshore facility under study in the FMEA and is responsible for getting Classification approval on it.

Redundancy. Ability of a component or system to maintain or restore its function when a single fault has occurred. Redundancy can be achieved for instance by installation of multiple components, systems, or alternative means of performing a function.

System Boundary. Imaginary perimeter encompassing all components within a specified system.

Validation Programme. See "Verification Programme."

Verification Programme. For the purposes of this book, the terms FMEA tests, FMEA proving trials, FMEA validation and FMEA verification programme are equivalent. They refer to trials and testing necessary to prove the conclusions of the FMEA or to establish conclusively the effects of failure modes that the FMEA desktop exercise had a high degree of uncertainty about.

Worst Case Failure (WCF). For dynamic positioning systems, the identified single fault in the DPS resulting in maximum effect on DP capability as determined through the FMEA study. This worst-case failure is to be used in the consequence analysis.

Worst Case Failure Design Intent. Minimum acceptable remaining system functionality after the worst case (most significant system impact) single failure. This is typically only applicable to DP system analysis.

Worst Case Failure Design Intent (WCFDI). For dynamic positioning systems, the worst-case failure design intent describes the minimum amount of propulsion and control equipment remaining operational following the worst-case failure. The worst-case failure design intent is used as the basis of design. This usually relates to the number of thrusters and generators that can simultaneously fail.

ACRONYMS AND ABBREVIATIONS

ABCU	Automatic Bridge Centralised Control Unmanned
ACC	Automatic Centralised Control
ACCU	Automatic Centralised Control Unmanned
ASOG	Activity Specific Operating Guidelines
BOP	Blowout Preventer
CDS	Certification of Drilling Systems
DFDE	Dual Fuel Diesel Engine
DG	Diesel Generator
DP	Dynamic Positioning
ER	Engine Room
ESD	Emergency Shutdown
EUC	Equipment under Control
FMEA	Failure Modes and Effects Analysis
FMECA	Failure Modes, Effects, and Criticality Analysis (normally pronounced "Fuh-Mee-Kah")
FO	Fuel Oil
HSE	Health, Safety, and the Environment
HVAC	Heating, Ventilation, and Air Conditioning
IACS	International Association of Classification Societies
IAS	Integrated Automation System
IEC	International Electrotechnical Commission
IL	Integrity level

IMCA	International Marine Contractors Association
IMO	International Maritime Organisation
ISQM	Integrated Software Quality Management Guide
MODU	Mobile Offshore Drilling Unit
OSV	Offshore Support Vessel
P&ID	Process and Instrumentation Diagram
PFD	Process Flow Diagram
PLC	Programmable Logic Controller
PMS	Power Management Systems
ROV	Remotely Operated Vehicle
SV	System Verification Guide
WSOG	Well Specific Operating Guidelines

Chapter 1

Introduction

BACKGROUND

In the marine and offshore industry, design and equipment configurations vary from one system to the next, and systems are in many cases increasingly complex. There are gaps in codes and standards which may lag technological innovations and there are issues related to interfaces between systems. Risk analyses such as Failure Modes and Effects Analysis (FMEA) provide a formalised approach to identify hazardous situations, address the gaps and interconnection variances, and improve safety, environmental performance, and operational downtime. Classes require clients to develop and submit FMEA as part of Classification requirements for certain systems and to obtain certain special notations. This book provides guidance and insight into the development process for FMEA to comply with Classification Rule requirements for various special notations. The utilisation of this guidance will provide tangible benefits as the marine and offshore industry is able to realise the positive results of FMEA that are developed correctly and managed appropriately throughout the lifecycle of a system. Some of these benefits include:

- FMEA that meet the intended objectives and are a support to the classification process.
- Consistency in scope, depth, and quality among comparable FMEA.
- Expedite the FMEA review process.
- Reduce failures, downtimes, and incidents.

PURPOSE OF THE FMEA

Whenever a system failure could result in undesirable consequences such as loss of propulsion, loss of propulsion control, etc., best practices advise conducting a risk analysis, such as an FMEA, as an integral part of the design and operational development process. This analysis can be a powerful aid in identifying possible failures which could potentially leave a vessel, an offshore installation, or its crew in peril. The ultimate goal of an FMEA from the point of view of Classification is to use it as supporting documentation to demonstrate compliance with the Class design philosophy and related Classification notation requirements and design intent for the particular system. There are instances where the goal of the vessel or asset owner is to have a comprehensive and systematic risk-based approach to the design. When such approach is taken, design choices are prioritised based on the assessment of risks, thus the much broader FMEA goal is to identify and reduce a wider range of risks that could arise from failures.

The ISQM (Integrated Software Quality Management) for software development is an example of such a risk-based design framework.

FMEA OVERVIEW

An FMEA is a design and engineering tool which analyses potential failure modes within a system to determine the impact of those failures. It was first developed by the US Department of Defense for use in systems design. The FMEA technique has since been adopted by commercial industries in an attempt to minimise failures and reduce safety, and environmental and economic impacts that could result from these failures. FMEA have more recently become a preferred risk analysis tool in the marine industry. It is required for certain systems by the International Maritime Organisation, by Classification Societies, select regulatory bodies, and industry groups to improve the safety of a design or operation, to increase its reliability and to minimise undesired events. As a risk management practice, FMEA are also an integral part of the design process of many proactive companies.

Summary of the FMEA Process

The FMEA is generated through a tabletop analytical process intended to identify system design and configuration weaknesses in all expected operational modes of the particular system. Once it has been determined that an FMEA will be performed and the scope of the study is agreed upon, an appropriate FMEA team of subject matter experts is assembled to conduct the analysis. A team is recommended for FMEA, in particular for larger systems requiring different specialties. In some instances, a study conducted by an FMEA practitioner knowledgeable in the system(s) being analysed and the development of FMEA is an adequate alternative. System boundaries are defined, and agreed upon, to clearly delineate what parts of the subject will be analysed. The team will include or interface with the owners/stakeholders to exchange data, including collection of system schematics, operational procedures and manuals and system configurations. The team brainstorms on the potential failure modes, their effects, detection methods and corrective actions. Recommendations are provided for corrective action throughout the development process and these recommendations may be ranked according to the severity of the potential effect.

This information is identified and recorded, usually in a tabulated format, and a preliminary report is issued to the owner/ stakeholders and team for review and verification of accuracy. An option is to recommend practical tests and trials to conclusively verify the analysis. For certain special notations and for certain organisations such as regulatory bodies, a further FMEA validation and trial programme must be developed and executed on the vessel in order to validate that the system responds to failures and failures are detected and alarmed as described within the FMEA. Once the comments from the team, owner and stakeholders on the preliminary document review have been received by the practitioner or FMEA team leader, the document will be updated and should be ready to be submitted to Class for review. The entity that has the contract with Class (e.g., shipyard, vessel owner) will have the ultimate responsibility for making sure the FMEA reports are submitted to Classification. The general elements of the FMEA process are discussed in detail in Chapters 1 through 6 and illustrated in Figure 1.1. Chapter 7 provides the specific guidance for selecting Class Classification FMEA requirements, as listed in Table 1.1 below. Faced with a particular FMEA requirement, the user may choose to go directly to the respective requirement in Chapter 7 for guidance and clarification.

Table 1.1 Index of System-Specific Guidance for Class FMEA Requirements

Class Vessel Rules; Offshore Support Vessels (OSV); Under 90 Metres; Mobile Offshore Drilling Units (MODU); Mobile Offshore Units (MOU); Offshore Facilities; High Speed Craft (HSC); High Speed Naval Craft (HSNC); Gas Fuelled Ships (GF); Propulsion Systems for LNG Carriers; Lifting Appliances

Automation
General Automation.
Computer-based Systems.
Wireless Data Communications for Vessel Systems.
Integrated Controls.
Electronically-controlled Diesel Engine.
Remote Control Propulsion.
Automated Centralised Control (ACC).
Automated Centralised Control Unmanned (ACCU).
Automated Bridge Centralised Control Unmanned (ABCU).
Gas Turbine Safety Systems.
Redundant Propulsion and Steering.
Single Pod Propulsion.
Dynamic Positioning Systems (DP)
Dynamic Positioning (DP) Systems
Integrated Software Quality Management (ISQM)
Software
Mobile Offshore Drilling Units (MODU)
Jacking and associated Systems
Offshore Support Vessels
Subsea Heavy Lifting
Certification of Drilling Systems
Drilling Systems/Subsystem/Equipment
Integrated Drilling Plant (HAZID)
Propulsion Systems for LNG Carriers
Dual Fuel Diesel Engine
Gas Fuelled Ships
Reliquefication, Dual Fuel Engine and Fuel Gas Supply
Lifting Appliances
Motion Compensation and Rope Tensioning Systems for Cranes

Fig 1.1 Process Flow for Classification Required FMEA

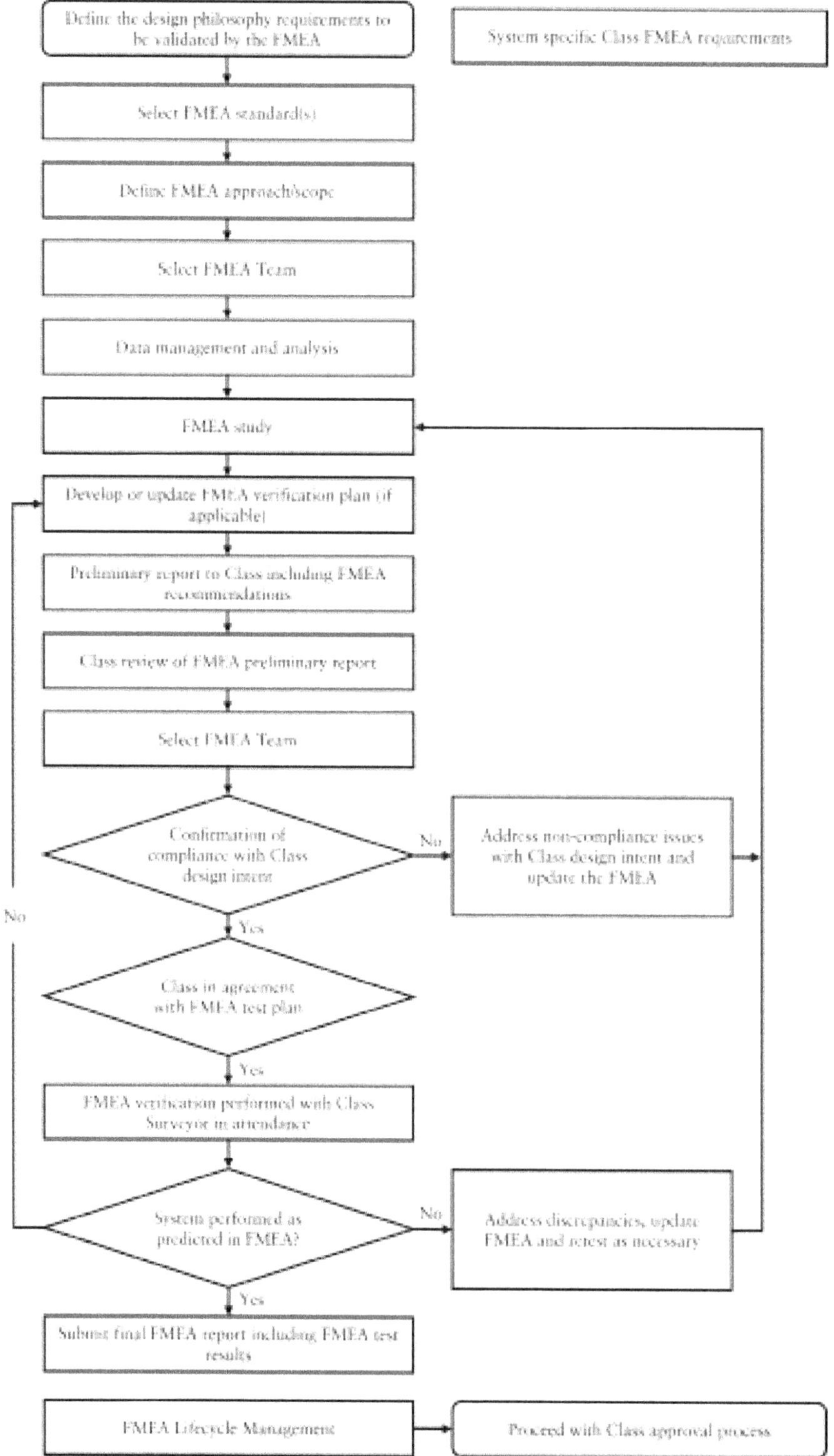

Chapter 2

Before the FMEA

PREPARING FOR THE FMEA

Conducting an FMEA or any risk analysis takes time, human resources, and funds. However, the best way to save on resources is to do a proper FMEA the first time. Poorly done FMEA take extra time and resources for revisions, corrections, and clarifications, and in many cases, repeated analyses. The following section provides an overview of the FMEA method, ground rules, assumptions, and constraints to take into consideration when performing an FMEA for Classification.

FMEA STANDARDS

By providing a clearly defined methodology and standards to be followed, the owner/stakeholder will be aware in advance how the FMEA will be generated and can have increased confidence in the results. Specifying standards does not guarantee an acceptable FMEA but it does guarantee an acceptable methodology and format. How well an analysis is performed, and to what level of detail, can only be achieved by selecting an FMEA team of subject matter experts or expert FMEA practitioner(s) experienced with the design, characteristics and performance of the systems being analysed, as well as someone knowledgeable in the technique to lead the analysis. Common FMEA standards used for reference include the following:

- IEC 60812, Analysis Techniques for System Reliability.
- US Military Standard MIL-STD-1629A, Procedures for Performing a Failure Mode, Effects and Criticality Analysis (cancelled in 1998 but it is still widely used as a reference).
- US Army Technical Manual TM 5-698-4, Failure Modes, Effects and Criticality Analysis (FMECA) For Command, Control, Communications, Computer, Intelligence, Surveillance, and Reconnaissance (C4ISR) Facilities, 2006.

There are several guidance documents that although developed for specific systems or types of vessels such as DP, High Speed Craft, or automation, provide a wealth of useful information that is applicable to other systems in the marine and offshore environment:

- IMCA M166, Guidelines for Failure Modes and Effects Analyses (FMEA).
- IMCA M178, FMEA Management.
- IMO MSC Circular 645, Guidelines for Vessels with DP Systems.

- IMO HSC Code, Annex 4, Procedures for failure mode and effects analysis.
- USCG MSC Guidelines for Qualitative Failure Analysis Procedure Number: E2-18
- USCG Marine Technical Notice 02-11, Review of Vital System Automation and Dynamic Positioning System Plans, refers to 46 CFR 62.20-3.

For FMEA for computer-based controls and software, the following general reference documents exist:

- National Aeronautics and Space Administration (NASA), "Software Safety Standard, NASA, Technical Standard, NASA-STD-8719.13B w/Change 1", July 2004.
- Nancy G. Leveson. "System Safety and Embedded Computing Systems" Aeronautics and Astronautics Engineering Systems, Massachusetts Institute of Technology (MIT), August 2006.
- Drs. Alex Deas, Sergei Malyutin, Vladimir Komarov, Sergei Pyko, Vladimir Davidov, "O.R. Rebreather Safety Case, FMECA Volume 5: Firmware and software", Revision A4, Deep Life Ltd., Glasgow, UK, August 2008.
- Haapanen Pentti, Helminen Atte, "Failure Mode and Effects Analysis of Software-Based Automation Systems," STUK, Radiation and Nuclear Safety Authority, Helsinki, Finland, STUK-YTO-TR 190, August 2002.

DESIGN PHILOSOPHY AND FMEA

It was noted earlier in this book that when Class requires FMEA, it is as a supporting document to verify that the system under review meets the specific Classification notation requirements and design philosophy. A general design philosophy for Classification is that a single failure shall not lead to an undesirable event or hazardous situation with immediate potential for injury to persons, damage to vessels, or pollution of the environment. Examples of these undesirable events can include loss of functionality of system (or degradation beyond an acceptable level) or loss of control of system. Only certain design solutions achieve the end result of avoiding the undesirable event. Corrective actions include:

- Redundancy in design.
- Safe and controlled shutdown and restart.
- Risk control diminishes the likelihood of the occurrence of undesired events.

Table 2.1 below shows the standard solutions for identified failures based on failure design philosophy and the undesired event.

Table 2.1 Typical Corrective Actions to Control Failure Scenarios

Undesired Events	*Solutions to comply with design philosophy that "No single failure shall lead to specified undesired event"*	*Example of Applicable Systems*
Any hazardous situation with immediate potential for injury, damage, or pollution	Safe and controlled shutdown	Most drilling systems (except those used for well control and active heave compensation) (Certification of Drilling Systems, CDS Notation)
Loss of Functionality of System[1]	Complete redundancy of system[2] Independent systems No common-cause failures[3]	Dynamic Positioning Systems (DPS-2 or DPS-3 Notation) Redundant Propulsion (R1 or R2 Notation) Redundant Steering (R1 or R2 Notation) Blowout Preventer (BOP)
Loss of System Control	Complete redundancy of controls and/or systems[2] No common systems or common cause failures	Computer-based systems Drilling Systems Controls (Certification of Drilling Systems, CDS Notation) ACC and ACCU systems

Notes:

1. *Loss of functionality or degradation beyond acceptable level.*
2. *Where complete duplication is not possible, robust, and reliable design that offers a proven low likelihood of failure may be accepted on a case-by-case basis. These non-redundant parts are to be further studied with consideration given to their reliability and mechanical protection. The details and results of these further studies are to be submitted to Class for review.*
3. *Definitions and examples of what can constitute common cause failure can be found in Chapter 3.*

FMEA SCOPE AND GROUND RULES

Although the basics of the FMEA technique are standard regardless of the system being analysed and the intent of the analysis, there is a certain level of customisation that depends on:

- Intent and scope based on Classification notation requirements being fulfilled.
- Type of system being analysed.
- Other goals of the owners/stakeholders.

The scope of a particular FMEA shall be defined at the outset of development and shall be agreed upon by the parties involved. Before the FMEA gets underway, the following scope items must be defined:

1. Physical and operational boundaries.
2. Failure criteria and types of failure.
3. Depth of analysis/level of indenture.
4. Design or operational philosophies (e.g., operating closed-bus vs. open-bus for a DP power distribution system.
5. What are the consequences of interest (undesirable events)?
6. Criticality ranking (FMECA) if desired.

Each of the items 1 through 6 above will be discussed in more detail in Chapter 2.

EQUIPMENT SCOPE AND PHYSICAL BOUNDARIES

The equipment to be analysed will be defined at the outset of the analysis based on the goal of the FMEA for Classification requirements and the type of equipment. For defining the physical boundary of the FMEA, it can help to answer the following questions:

*   What are the main systems/subsystems/equipment of interest in this FMEA?
*   Systems/equipment interfacing with the main system under study?
*   Supporting utilities?
*   Control systems?
*   What is excluded from the FMEA?

Table 2.2 provides an example of the physical boundaries and the equipment that will be subjected to analysis for a Dynamic Positioning FMEA to comply with rule requirements. All systems that have functional or physical interfaces with the system under study should be identified and should be subject to consideration in the FMEA. Examples of interfaces are data/signal communication between systems, input, and outputs between systems, and even layout issues may provide interdependence that need to be considered. As a minimum, the failures at the interfaces should be postulated and analysed in the FMEA. For more detail, refer to IEC60812, Section 5.2.2.2 – Defining system boundary for the analysis.

Table 2.2 Examples of System/Subsystem's Physical Boundaries (for a DP System)

System		Subsystem		Components
1	Marine Auxiliary Systems	1.1	Fuel	
		1.2	Remote Control Valve	
		1.3	Engine and Generator Lubricating Oil	
		1.4	Seawater Cooling	Each subsystem has components whose failures should be evaluated. Representative component lists for select subsystems are included herein.
		1.5	Freshwater Cooling	
		1.6	Charge Air	
		1.7	Compressed Air	
		1.8	Emergency Generator	
		1.9	Engine Management and Safety System	
		1.10	HVAC and Chilled Water System	
2	Power System	2.1	Generators	
		2.2	Power Distribution	
3	Propellers and Steering Gear	3.1	Main Propellers	
		3.2	Steering Gear	Power supply, main, auxiliary and backup pumps, hydraulics, cooling, controls (main, alternative, emergency), control power supply, protections, angle indications, alarms, ready signals, etc.
		3.3	Tunnel Thrusters	
		3.4	Azimuth Thrusters	

Table 2.2 Examples of System/Subsystem's Physical Boundaries (for a DP System) (continued)

System		Subsystem		Components
4	Vessel Management System	4.1	Switchboard PLC Communications	Each subsystem has components whose failures should be evaluated. Representative component lists for select subsystems are included herein.
		4.2	Air Conditioning	
		4.3	Network Topology	
5	DP Control Systems	5.1	Independent Joystick (IJS)	
		5.2	Power Management System (PMS)	
		5.3	Networks	
		5.4	Reference and Sensors	DP control system and interfaces (including position reference systems, gyros, vertical reference sensors and wind sensors). Interfaces can include data interface with tensioner system, interface with survey package, etc.
		5.5	DP Alert and Communications	
		5.6	Cable Routes	
		5.7	Backup DP Systems	
6	Emergency Shutdowns	6.1	Emergency Shutdown System (ESD)	
		6.2	Fire and Gas	
		6.3	Thruster Emergency Stops	
		6.4	Group Emergency Stops	Each subsystem has components whose failures should be evaluated. Representative component lists for select subsystems are included herein.
		6.5	Fans	
		6.6	Dampers	
		6.7	Firefighting Systems	
		6.8	Quick Closing Valves	

Table 2.2 Examples of System/Subsystem's Physical Boundaries (for a DP System) (continued)

System	Subsystem		Components
	7.1	Riser (MODU)	
	7.2	Tensioner (Pipe/product lay vessel)	
	7.3	Crane	

Source: *Modified from the Marine Technology Society DP Committee, Technical and Operational Guidance TECHOP ODOP O4 (D) FMEA Gap Analysis, October 2013*

OPERATIONAL BOUNDARIES (GLOBAL AND LOCAL)

Global operations are the overall operations of the facility or vessel. Local operations are the operations of the system that is within the boundaries of the FMEA. Each operation and operational combination can present distinct failure scenarios, (failure modes, hazards, and consequences). The FMEA must define the global operations at the installation or vessel level as well as the local operations of the system that is within the boundaries of the FMEA. All operational combinations should be considered in the FMEA, as well as the switching between modes. Examples of global operations for a DP vessel may include station keeping; weathervaning; ROV following; manoeuvring; pipe-laying (if a product-laying offshore support vessel); drilling (if a MODU); and operating underway and fully laden for a tanker. Examples of local operational modes for the DP system would include power management systems (PMS) configurations; blackout recovery; power load shed; manual; joystick; independent joystick system (IJS); and switching between modes.

FAILURE CRITERIA AND TYPES OF FAILURE

The proper and comprehensive identification of failures is a fundamental step in the FMEA exercise. Several considerations and basic assumptions regarding the failures to be considered should be understood by all team members before starting the FMEA. These include single failure criteria; hidden failures; common cause failures; treatment of unavailable systems; failure of passive and active components; and consideration of external events. Further details on each of these are provided throughout Chapter 3.

DEPTH OF ANALYSIS

The equipment level to which an analysis is performed is very much dependent upon the system being evaluated, the intent of the FMEA and the goals of the owner/stakeholder. In general, FMEA for the marine industry do not attempt to identify every possible fault of every component in the system but will proceed to a level where additional analysis of failure modes from lower-level components will not reveal additional effects on the system. Finding the appropriate level at which to stop on a given system is somewhat of an art and is developed with experience. An experienced FMEA team/practitioner should be able to determine the optimal depth of analysis that is sufficient to satisfy the intent of the FMEA. The System-Specific FMEA Requirements tables presented in the body of this book give a more definite guideline of the depth of the analysis sought for most FMEA requirements for Classification.

    Example: defining the depth of the analysis. Let us consider a failure analysis on an automation system. The analysis typically would be performed assessing the failure for the acquisition unit modules, failure of inputs and outputs. A typical FMEA will stop at the module card level. Performing an analysis of individual circuit boards, resistors or failed data paths within the computer would not necessarily contribute to an assessment of the system as a whole. The end effect of those lower-level failures on the system would be the same as failures already assessed at the higher level, therefore the recommendations to recover from those failures would already be addressed. Above example refers to a standard automation FMEA which assumes the code is being programmed correctly. Note that the assumption may be different for ISQM software-focused FMEA.

For more detail, refer to IEC60812, Section 5.2.2.3 – Levels of analysis.

CRITICALITY RANKING (FMECA)

Failure Modes, Effects, and Criticality Analysis (FMECA) is an extension of the FMEA process which includes an additional criticality assessment. Criticality ranking explicitly and transparently brings to prominence the most critical issues and is extremely helpful for deciding the corrective actions. In the development, follow-up and implementation process of corrective actions, the criticality ranking helps to evaluate that the effort, time, and resources are commensurate with the criticality of the item. Criticality rankings based on risk use a combination of the consequence (severity) of the failure and the anticipated likelihood of the consequence occurring. The analysis will highlight failure modes with high probability of occurrence and severity of consequences, allowing corrective actions to be implemented where they will produce the greatest impact. Ideally, frequency estimates will be based on historically quantifiable data from the field, but in many cases data of this type is unavailable or poorly documented. The methods to determine criticality must be clearly defined prior to embarking on an analysis since the veracity of failure data (or lack thereof) can greatly influence the results.

It is worth noting that some standards make a difference between a qualitative and quantitative criticality assessment. The quantitative assessment as described by the MIL STD 1629 is quite involved and will not be discussed here as it not a requirement of Classification. The qualitative assessment of criticality uses expert judgment to place the event in criticality or risk matrix or a criticality benchmark. When a criticality analysis is asked for by the Classification requirement, the qualitative criticality assessment is not only sufficient, but also the recommended method. The most common method for qualitative evaluating the criticality is the use of ranking systems which scale severity of consequence vs. likelihood, as shown in Figure 2.1. The matrix in Figure 2.2 example has four levels of consequence and four levels of likelihood. More levels can be defined as needed, but anything less than four levels may not provide enough granularity to make appropriate risk-based decisions. Given the overall lack of reliability data for many marine systems and components, performing an assessment on a qualitative level based on experience and knowledge of the system under study is sometimes the only means by which to achieve a meaningful criticality assessment. A high severity and high likelihood event are not acceptable, and risk control measures to reduce either the likelihood of occurrence or severity may need to be developed.

Very few Classification requirements specifically request an FMECA. As detailed in Chapter 7, ISQM and System Verification do explicitly require an FMECA. However, Classification will accept any voluntary submission of an FMECA instead of an FMEA.

FMEA NAMING CONVENTION WITHIN THIS BOOK

An FMECA is an FMEA with an additional analysis of the criticality of each failure scenario. Thus, the term "FMEA" is used generically in this book to include both FMEA and FMECA. For example, the next section discusses the "FMEA Team," and it could be entitled "FMEA/ FMECA Team" since its content is equally applicable to FMECA. However, it would be tiring to the reader to see the convention "FMEA/FMECA" throughout this book. The reader can assume that anytime the term FMEA is used in this book, it can be substituted by the term FMECA. This inverse, however, does not hold true. When the term FMECA is used, it can only mean an FMEA with the criticality extension.

Fig 2.1 Typical Risk Matrix for FMECA

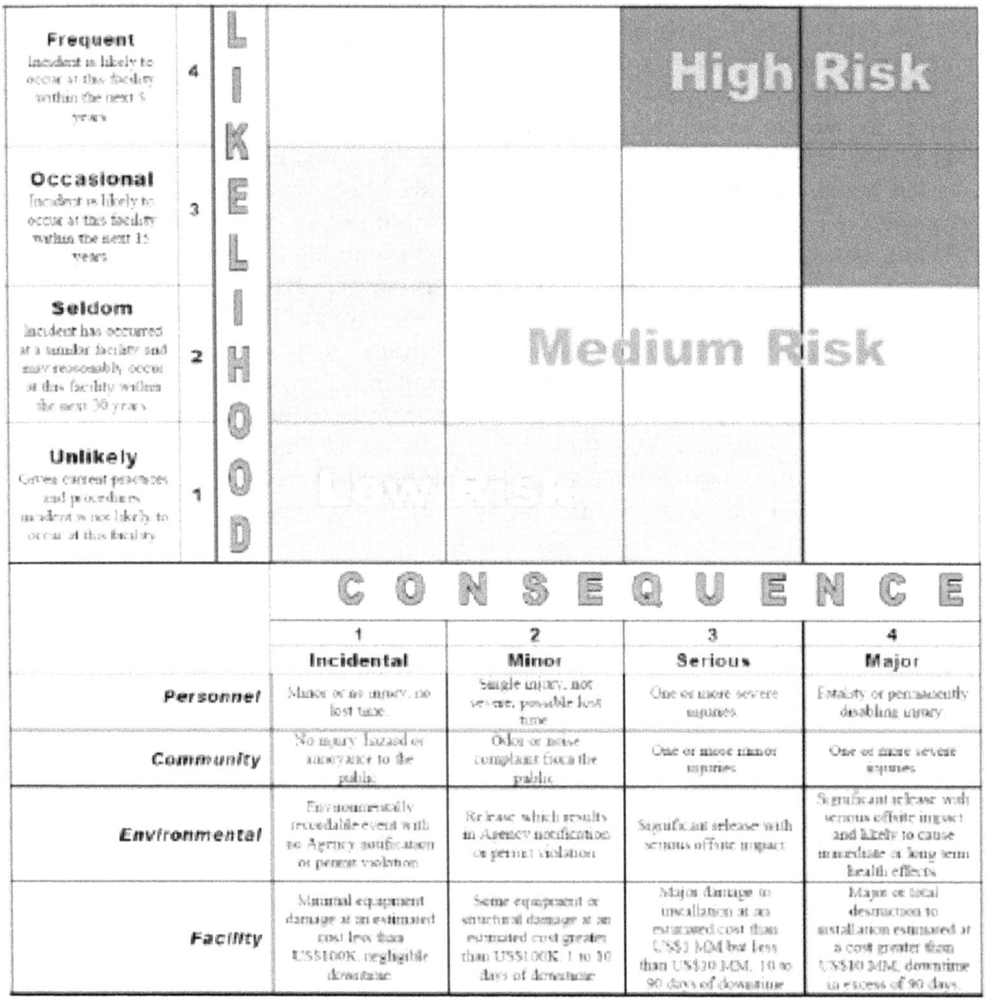

LIKELIHOOD

Frequent	4	Incident is likely to occur at this facility within the next 5 years
Occasional	3	Incident is likely to occur at this facility within the next 15 years
Seldom	2	Incident has occurred at a similar facility and may reasonably occur at this facility within the next 30 years
Unlikely	1	Given current practices and procedures, incident is not likely to occur at this facility

High Risk — Medium Risk — Low Risk

CONSEQUENCE

	1	2	3	4
	Incidental	**Minor**	**Serious**	**Major**
Personnel	Minor or no injury, no lost time	Single injury, not severe, possible lost time	One or more severe injuries	Fatality or permanently disabling injury
Community	No injury hazard or annoyance to the public	Odor or noise complaint from the public	One or more minor injuries	One or more severe injuries
Environmental	Environmentally recordable event with no Agency notification or permit violation	Release which results in Agency notification or permit violation	Significant release with serious offsite impact	Significant release with serious offsite impact and likely to cause immediate or long term health effects
Facility	Minimal equipment damage at an estimated cost less than US$100K, negligible downtime	Some equipment or structural damage at an estimated cost greater than US$100K, 1 to 10 days of downtime	Major damage to installation at an estimated cost than US$1 MM but less than US$10 MM, 10 to 90 days of downtime	Major or total destruction to installation estimated at a cost greater than US$10 MM, downtime in excess of 90 days

UNITED STATES COAST GUARD SUPPLEMENTAL REQUIREMENTS FOR QUALITATIVE FAILURE ANALYSES (QFA)

Vessels sailing under the flag of the United States require an FMEA or "qualitative failure analysis" for many of the same systems for which Classification requires FMEA. As per 46 Code of Federal Regulations "one copy of a qualitative failure analysis must be submitted for propulsion controls, microprocessor-based system hardware, safety controls, automated electric power management, automation required to be independent that is not physically separate and any other automation that in the judgment of the reviewing authority potentially constitutes a safety hazard to the vessel or personnel in case of failure. The QFA should enable the designer to eliminate single points of failure." The qualitative failure analysis is intended to assist in evaluating the safety and reliability of the design. It should be conducted to a level of detail necessary to demonstrate compliance with applicable requirements and should follow standard qualitative analysis procedures. The QFA must be explicitly listed:

- Assumptions.
- Operating conditions considered.
- Failures considered.
- Cause and effect relationships.
- How the crew detects failures.
- Alternatives available to the crew.
- Necessary design verification tests should be included.

Questions regarding failure analysis should be referred to the reviewing authority at an early stage of design.

FMEA TEAM

There are two typical styles of conducting an FMEA. One is using a workshop setting with in-house subject matter experts with first-hand knowledge on the system being analysed. The other style is to hire a third-party FMEA practitioner to develop the FMEA. The FMEA practitioner assembles a multidisciplinary team to perform the analysis, which is developed nearly independently from the stakeholders, other than for initial inputs and review of the FMEA. The appropriate multidisciplinary FMEA team is selected based on specialised knowledge needed for the analysis. The disciplines that form the FMEA team should include subject matter experts in machinery, control, electrical and naval architecture, as applicable. The team should also have knowledge of design, manufacturing, assembly, service, quality, reliability, and operation. A workshop-type FMEA must have an FMEA facilitator, who is primarily:

1. Knowledgeable of the FMEA technique.
2. Has effective communication and administration skills.
3. Familiar with the type of system to be analysed and its intended operation.

STAKEHOLDER WORKSHOP SETTING

Stakeholders' multidiscipline workshop FMEA are developed using a meeting setting where several parties are directly and simultaneously involved in the process. Participants might include the builder/ shipyard, third-party FMEA practitioners, Classification, operators, owners, etc. Workshops are very useful when applied early in the design stage of a system since the analysis will involve several parties with a vested interest in the outcome of the design and takes place at a time when the design can be modified if required. A potential downside to forming a workshop of this nature is that some of the parties involved may have conflicting interests, which can slow the process down or impede free sharing of information needed to develop the FMEA. The workshops are simple brainstorming sessions conducted in enough detail to identify and discuss specific failure modes. Part-time participants may be commissioned to provide input according to their area of expertise, usually via part-time participation in the FMEA workshop.

THIRD-PARTY FMEA PRACTITIONER(S)

A third-party FMEA practitioner working solo or a team of FMEA practitioners may be contracted to perform the FMEA. The level of interaction between the team and the stakeholders, and within the team itself will depend largely on the scope of the analysis, specific team member expertise and experience and size of the project. The practitioner team may perform the analysis in a workshop style where the team meets to examine the system under study. They may also perform the analysis in a more independent fashion where the team initially meets for a collaborative brainstorming session to pool data and concepts for the analysis, each member then independently performing an analysis in their respective areas of expertise and then reconvening for a group review of the overall analysis. The analysis may also be performed without workshops or brainstorming sessions. Team members may already be familiar with the type of system being analysed and may be able to proceed into individual analyses without requiring a workshop or brainstorming session. Each piece of the analysis would then be integrated by a project lead with oversight of the entire project. Where the scope of analysis is limited or for simpler systems, an individual with the appropriate level of knowledge and experience may be sufficient to perform the FMEA. Though only a single person is performing the analysis, the overall development process remains the same.

Example: typical team selection. In the case of an FMEA for a Dynamic Positioning (DP) vessel, a typical third-party practitioner team consists of individuals with expertise in mechanical, electrical, DP operations as well as an individual familiar with all systems involved to take leadership of the FMEA and provide an integration function between the various system experts. For a stakeholders multidiscipline team conducting the FMEA in a workshop format, the disciplines will be the same as in the third-party practitioner team but will also need an FMEA workshop leader/facilitator with experience in the FMEA technique and a technical recorder to capture all the information generated during the workshop.

CLASS PARTICIPATION IN THE FMEA WORKSHOP

As indicated in Chapter 2, the FMEA can be conducted as a stakeholders workshop setting or commissioned to a third party. The stakeholders workshop style is the preferred method to take into account the opinions and experience of all the stakeholders. There is no requirement that

Class personnel be part of the FMEA workshop. However, benefits can be derived from the participation of a Class Engineering representative that will be directly involved in reviewing the FMEA and the system in order to grant Classification approval. Some of the benefits include:

1. As a participant in the FMEA workshop, the Class Engineering representative will be able to point out the issues that Class considers relevant for the classification of the proposed design, and thus should be discussed during the FMEA; and
2. Participation in the FMEA workshop of the Class representative that will be reviewing the FMEA will minimise the number of questions and clarifications at the time of the Class review of the FMEA because he/she will be familiar with the study and design.

TEAM PREPARATION

Team preparation should include project introduction and discussion on the scope of the analysis. Any training needs should be identified. For example, with many FMEA facilitation tools readily available, is the technical recorder adequately trained to use the selected tool? If specific fault identifying techniques are going to be used in the process of identifying failure modes particular to the analysis, is the team capable of applying them? Such analysis techniques may include root cause analysis, fault tree analysis, etc. The team should also be clear on the approach, ground rules, assumptions and constraints placed on the analysis and be able to appropriately address those limitations.

IDEAL TIMING TO CONDUCT FMEA

Though an FMEA can be conducted at any point in the lifetime of a system, it is most advantageous to be performed early on in the design process. FMEA performed early in the design stage have the advantage of catching design or system configuration issues in time to allow for modifications before construction. The FMEA can be used as an input in the design review process and will be updated to reflect any design changes. Although fully integrating an analysis of this type into the design process can slow development and increase costs due to design changes, the benefits outweigh the initial expenditure. The benefits include:

1. Safer design as the most optimum risk control options are typically those incorporated early in the design process; and
2. The savings realised by eliminating costly retrofits or system upgrades post-build may outweigh the initial expenditure.

The usefulness of an FMEA will largely depend on the level to which its findings and corrective actions were incorporated in the design process. The timing and delivery for an FMEA performed to comply with Classification requirements will depend largely on the type of FMEA being performed and the requirements that need to be satisfied. Small self-contained systems can have their FMEA performed at the vendor stage. Larger systems that are heavily integrated with other systems should have an FMEA performed by the system integrator prior to installation/ commissioning. In some cases, it may be necessary to have an FMEA of the small self-contained system early on, and then the larger-scope integration FMEA (or similar type of risk study) later on in the design. Regardless of when the FMEA study takes place, Classification society reviewers would look for evidence (e.g., updated drawings, documentation, etc.) that the FMEA findings were made part of the design and integration process.

For a new vessel or installation, the building contract should specify that input from the FMEA or risk study requested by Classification be taken into consideration, and the FMEA should be commissioned as early as possible in the project. This needs to be negotiated accordingly in the contract.

For an existing vessel or installation, company management needs to be aware that changes to the system may be required as a result of the recommendations arising from the FMEA and that sufficient funds must be made available to meet the changes. FMEA should be updated or re-performed whole when modifications are made to an existing system covered by the FMEA or when retrofits or replacement controls systems are installed. The extent to which the FMEA is modified will depend upon the nature of the changes to the system(s) being analysed. On the other hand, if a vessel is converted for new purposes and it is required to have an FMEA, the output of the FMEA may be limited to the systems affected by the changes. Although extremely useful in discovering potential issues related to failure modes and their effects, FMEA performed on existing systems have the disadvantage of not being integral to the system design process and FMEA recommendations may drive costly system modifications to meet the established criteria. If the FMEA is to be performed on an existing system, the solutions to FMEA findings can be managed through system upgrades or mitigated by operational procedures.

Chapter 3

Developing the FMEA

DEVELOPING THE FMEA

The following sections describe the development steps typically followed when performing an FMEA:

1.   Data Management.
2.   FMEA Study; and
3.   FMEA Report.

DATA MANAGEMENT

Data Collection to Support the Analysis

After concluding the initial tasks of establishing a scope and intent of study, selecting a team, etc., the owner/stakeholder will be engaged to deliver as much information on the subject of the FMEA as possible. This information might include general configuration and layout data, hardware listings, system schematics (such as electrical, HVAC, piping diagrams, etc.), previously performed system/subsystem FMEA, prior trials documentation and operational philosophy documentation. Vendor-specific FMEA may also be referenced for each piece of equipment but are typically generalised documents that do not include installation specifications required for an accurate system analysis. Data may also be collected by interviewing design personnel, operations, testing, and maintenance personnel, component suppliers and outside experts to gather as much information as possible.

Other Risk Analysis as Input to the FMEA

Other types of risk analysis such as HAZIDs and HAZOPs are sometimes used for input in the preparation of an FMEA. They can play a significant role in gathering data for failure analysis, particularly for systems without historical data to review or previous analyses. Other tools that might be used or fed into the process include fault tree, root cause and event tree analyses. They are described in detail in Table 3.1. Note that none of these items are a standard input to FMEA development, but, if available, they can provide invaluable input, in particular to an FMEA that is being developed by third-party FMEA practitioners who more than likely were not involved in previous design or analysis activities for that particular vessel/system.

Data Analysis

The data analysis process used by the FMEA team is iterative and incorporates the above data sources, as available, to develop an overall analytical approach to the FMEA. The FMEA team facilitator will perform an assessment as data is received to determine if additional information is needed on system design, operational concepts, procedures, etc., and engage the necessary parties (including hardware vendors) to obtain the desired information and subsequent distribution to relevant FMEA team members. By approaching data collection and analysis in this manner, the team can establish a complete picture of the subject and concurrently focus on areas of interest to identify potential issues more readily.

FMEA STUDY

The overall flow of the FMEA process is outlined in Figure 3.1 below. Each of the blocks is discussed in detail in subsequent pages.

Defining the Analysis

The system to be analysed is defined by the system physical and operational boundaries, the equipment scope and depth of the analysis, system functions, interface functions, expected system performance and constraints, and failure definitions. High-level functional narratives of the system including descriptions of tasks to be performed for different operational phases and modes should also be developed at this stage.

DEVELOPING THE ANALYTICAL APPROACH

The most common method for analysis to identify failure modes is the use of failure mode worksheets supported by functional/reliability block diagrams.

Table 3.1 Risk Analyses that could Provide Input Information to an FMEA

Types of Analysis	*Description*
HAZID	Hazard Identification (HAZID) is a process used to examine potential causes of hazardous events (accidents), how likely the stated event might occur, what the potential consequences are if it did, and what options there are for preventing/mitigating the event. Hazard identification is an integral part of the risk assessment and management process. A HAZID study is performed via brainstorming workshops consisting of individuals with expert knowledge of the systems under study. The team identifies and classifies hazards using checklists, "what-if" lists, accident, and failure statistics, by experience from previous projects, etc.
HAZOP	A Hazard and Operability study (HAZOP) is a qualitative, structured, and systematic examination of processes and operations to identify and evaluate issues that may represent risks to personnel or equipment or prevent efficient operation. The HAZOP technique was initially developed to analyse chemical process systems but has been extended to other types of systems and complex operations. Much like a HAZID study, a HAZOP study is performed via workshops where experts meet to examine the system under study. The method can be applied to any process where design information is available. This commonly includes a process flow diagram which is examined in small sections, such as individual components and common equipment. A design intention is specified for each of these. The HAZOP team then determines what the possible significant deviations are from each intention, along with feasible causes and likely consequences. It can then be decided if existing safeguards are sufficient or if additional actions are necessary to reduce risk to an acceptable level.
Root Cause Analysis	Root cause analysis is a process designed for use in investigating and categorising the fundamental cause of an initiating event with safety, health, environmental, quality, reliability, and production impacts. The analysis helps identify what, how and why something happened with the intent of developing recommendations for corrective measures to prevent future occurrences of similar events.
Event Tree Analysis	Event tree analysis is a technique to identify and evaluate the sequence of events that results from an initiating event. The analysis is performed by creating "event trees" that follow a logical sequence. An event tree analysis can result in different possible outcomes from a single initiating event. The objective of the analysis is to determine whether an initiating event will result in a mishap or if the event is sufficiently controlled by safety systems or procedures.
Fault Tree Analysis	Fault tree analysis is a top-down, deductive failure analysis in which an undesired state of a system is analysed using Boolean logic to combine a series of lower-level events. This analysis method is mainly used to determine the probability of an accident or a particular functional failure.

Table 3.1 Risk Analyses that could Provide Input Information to an FMEA (continued)

Types of Analysis	Description
Bow-Tie Analysis	Bow-Tie analysis is a risk approach that graphically displays the relationships between hazardous events, its causes and consequences and the risk control barriers in place to stop the accident sequence. Bow-Ties can be used as a simplified single diagram joining for an undesired event, its fault and event trees into one visualisation. The left side of the Bow-Tie is a pseudo fault tree that seeks all the potential precursors (or failures) and underlying causes of the accident and preventive control measures to avoid it. The right side of the Bow-Tie is a pseudo-event tree that describes the potential outcomes of a top event, and what controls are in place to prevent or mitigate the outcomes.

** Analyses marked with * are conducted for a select number of scenarios, where more detailed analysis is needed regarding the types of failures, combinations of failures, causes, potential outcomes etc. They are likely to be conducted after the FMEA to shed light into scenarios that had uncertainties.*

Fig 3.1 FMEA Study Flowchart

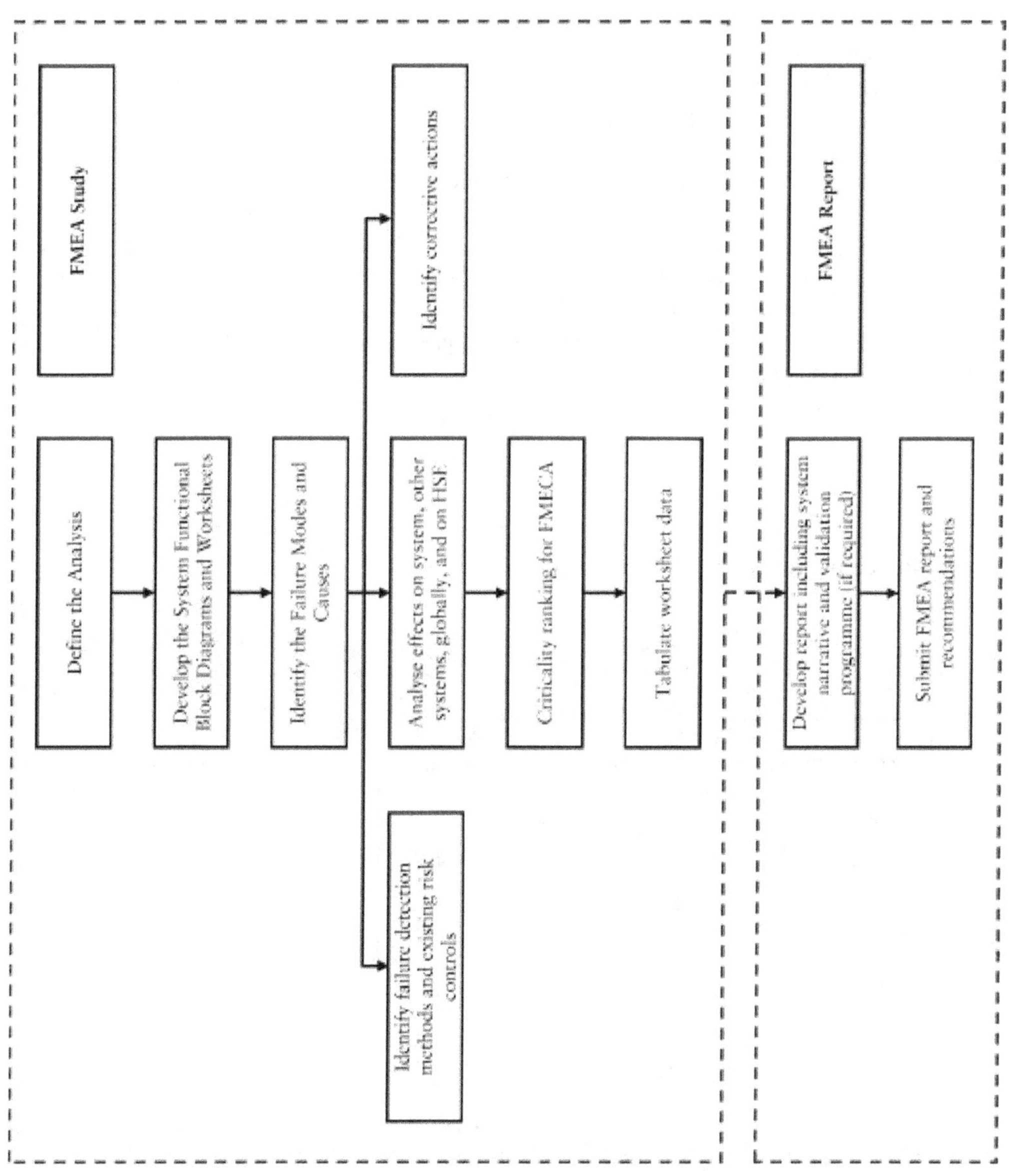

FUNCTIONAL BLOCK DIAGRAMS AND RELIABILITY BLOCK DIAGRAMS

A functional block diagram is used to show how the various parts of the system interact with one another. They are powerful illustrative tools which aid in visualising the interfaces and interdependencies between elements involved in system functionality. Block diagrams provide:

1. A high-level basis by identifying the chain of required systems/subsystems needed for successful operation.
2. For easier identification of possible failure modes and effects, failure causes and possible locations of hidden failures.

A special type of block diagram, the reliability block diagram (or dependence diagram), is well-suited for aiding in defining the scope and physical boundaries of an FMEA, in particular, FMEA to prove the redundancy of a system. A reliability block diagram represents a system with its major parts drawn as blocks connected to each other, either in series or in parallel (refer to Figure 3.2). A path in series indicates that if any of those blocks fail, the entire system fails. The parallel paths indicate there is a redundancy for that particular block and the system as a whole can still continue to operate through the other parallel path.

Fig 3.2 Reliability Block Diagram (or Dependency Diagrams)

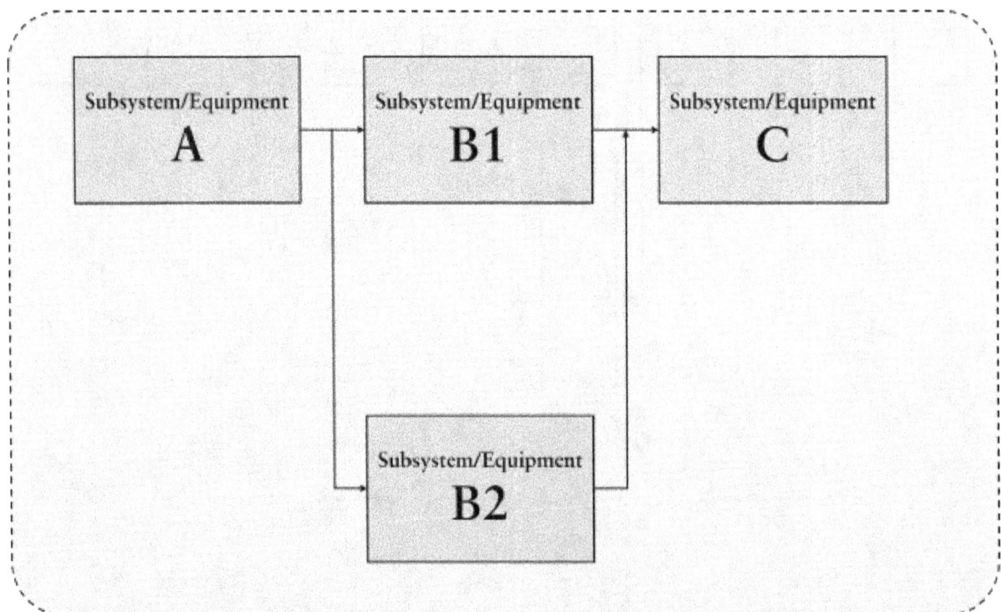

One method for analysing the system being studied is to break it down into various levels (i.e., system, subsystem, equipment, and field replaceable components). Schematics and other engineering drawings are reviewed to show how different subsystems, equipment, or components interface with one another by their critical support systems such as power, hydraulics, actuation signals, data flow, etc., to understand the normal functional flow requirements. A list of all functions of the equipment can be prepared before examining the potential failure modes of each of those functions. Operating conditions such as temperature, loads, and pressure, as well as environmental conditions may be included. The general vessel/installation operations should be understood and the likely impact of failure within the operating environment should be related to the analysis process. This impact can be illustrated with a pipe-laying activity. Figure 3.3 shows how the pipe-laying operation has an effect on the positioning of the vessel by applying a force to the DP system by virtue of the weight and size of the pipe being laid. This force is a function of the depth of water and size of pipe being laid and is thus project specific. The FMEA study of the pipe-laying tensioner should include the potential wider impact on the vessel. If the pipe tensioner stops operating, the vessel is effectively anchored to the seabed by the pipe. This could be relevant to worsening weather conditions.

Fig 3.3 Example of external/operational forces that may impact the FMEA Study

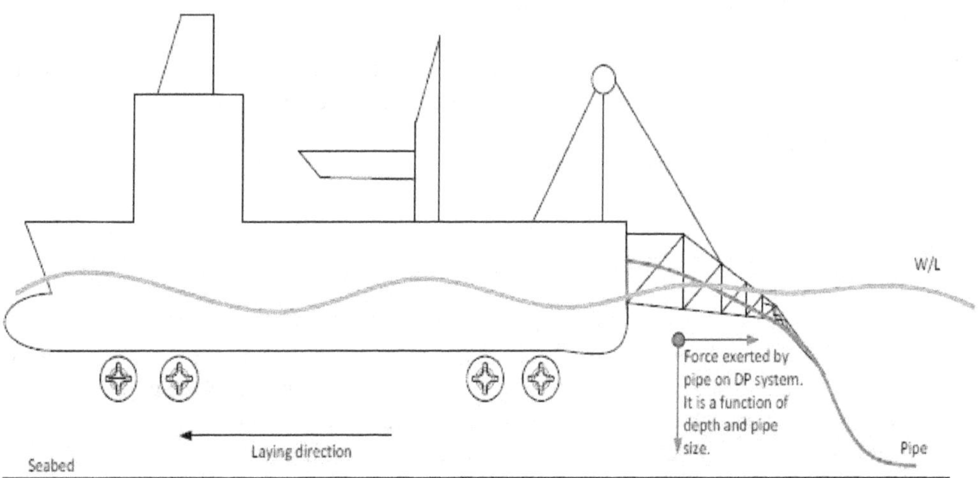

Failure Mode Worksheets

FMEA worksheets are tabulated data related to the identified failure modes and are a necessary organisational tool for performing the analysis that allows a condensed reference to all pertinent aspects of the analysis. A typical worksheet is populated with numerous pieces of information including the item or function being analysed, associated failure modes and causes, local and global effects on the system, detection methods, safeguards, recommended corrective actions and risk/criticality classification. There is no set standard or style required for a worksheet, so content and organisation may vary. However, it is important that content is sufficient to capture the analysis and convey relevant information regarding the system and the intent of the analysis. An example of a blank FMEA worksheet is provided in Table 3.2. Note the highlighted columns for severity and likelihood assessments. This is risk-based ranking or criticality for the particular scenario that is captured during the analysis, thus turning the FMEA into a Failure Modes, Effects, and Criticality Analysis (FMECA). This is an optional but especially useful step to help prioritise corrective actions.

IDENTIFY FAILURE MODES

Potential failure modes are determined by studying the relevant data compiled, in particular the functional element outputs. Failure modes within the physical and operational limits of the study scope are identified and described. The following typical failures are to be considered when determining failure modes and causes, but are by no means a complete list:

- Premature or spurious operation.
- Failure to operate when required.
- Intermittent operation.
- Failure to stop operating when required.
- Loss of output or failure during operation; and
- Degraded output or degraded operational capability.

Table 3.3 offers an example of failure modes used for FMEA equipment, including the equipment control system. Note that these lists of failure modes are only examples. Different lists would be required for diverse types of systems. In addition, there are certain ground rules and considerations regarding failures that need to be in the mind of the FMEA participants, such as:

- What types of failure shall be discussed – functional failures vs. component-level failures?
- If multiple failures occur simultaneously in order to result in undesirable events, shall such a scenario be analysed?
- What if the failures did not occur simultaneously but one failure could have occurred and was not detected, such as in safety equipment, and only when there was a demand for the safety equipment as a result of another failure, the safety equipment failure was discovered?
- How might a common initiating result in simultaneous failures of equipment? For example, the loss of power supply can result in the loss of function of multiple equipment.

These considerations are discussed in the following sections. Also, refer to IEC 60812, Section 5.2.3 for further guidance on Failure Mode Determination.

Table 3.3 Sample Failure Modes of Mechanical and Electrical Components

Mechanical Components	*Electrical Components*
External Leak or Rupture	Loss of/Degraded Power
Internal Leak	Fails with No Output Signal or Communication
Plugged	Fails with Low or High Output Signal
Mechanical Failure (e.g., fracture, galling, fatigue)	Erratic Output
Mechanical Damage (e.g., broken by external forces)	Fails to respond to Input
Wear	Processing Error
Corrosion or Erosion	Electrical Short
Loss of Function	Loss of Function
Loss of Pressure	

Functional Failures/Functional FMEA

A common approach for the FMEA is to analyse failures related to a particular function of the equipment not being performed or being performed incorrectly. Let us assume a system needs to pump x gpm from point A to point B. Typical functional failures for such a system would include failure of pumping capability, pumping at a rate below requirements, pumping at a rate exceeding requirements and pumping backwards. The causes or failure mechanisms for these functional failures would include motor failure; loss of power; degraded pump or motor, under voltage to motor; over voltage to motor; leaky non-return valve on discharge of pump. In order to perform a functional failure FMEA, the functions of the item under review must be defined. Note that the system/equipment under review may have more than one function.

Single Failure Criteria

FMEA for Classification are typically performed to assess single failures and their effects (i.e., two simultaneous independent failures are not considered). It is customary to also consider a single act of maloperation as a single failure. Assessments of this type are usually limited to errors that would result in unwanted consequences. A "single act" is generally taken to mean the operation of a single button, switch, lever, etc. There are two distinct instances when more than one failure should be considered in the FMEA:

1. When one of the failures can be latent, undetected, or hidden.
2. When two or more systems or components can fail due to a single specific event or cause (common cause failures).

Both of these exceptions to the "single failure" criteria are discussed in the next section.

Hidden Failures

An exception to the single failure criteria is for the case of latent, or hidden, failures where their presence is undetectable. In such cases, single failures in combination with an initial hidden failure and their combined consequences will be analysed. Since the initial hidden failure is unknown until the second failure occurs, the two failures will be considered together as a single event. Equipment that performs a back-up function and is in a non-operational or standby state may fall into this category if the functionality of the stand-by equipment cannot be verified until it is activated. Likewise, most safeguards and barriers are prone to hidden failures. They are not needed for operation and without proper monitoring to detect their failure. It is only discovered when there is a demand for the safeguard due to another failure. Refer to Chapter 3 for special considerations on hidden failures of safeguards and barriers. It is important to note that not every hidden failure will necessarily be assessed. The level to which hidden failures are assessed depends on the goals and intent of the analysis. Typically, the analysis will be limited to hidden failure/additional failure combinations that lead to an undesired event, but loss of either component on its own will not.

Example: hidden failures. Take a dual pump system where the loss of both pumps is considered unacceptable. Switch-over from the primary to the back-up pump is performed automatically by a separate controller (which does not have fault detection). A failure of the switch-over controller (undetected) followed by a failure of the primary pump would cause a total loss of the system since switch-over to the secondary pump would not occur. In this case, the two failures would be considered a single event since the first failure (controller) was unknown until the second failure (primary pump) was realised.

Common Cause Failures

A common cause failure is the loss of two or more systems or components due to a single specific event or cause: a design deficiency, a manufacturing defect, operation and maintenance errors, an environmental issue, an operator-induced event, or an unintended cascading effect from any other operation, failure within the system, or a change in environmental conditions. For the purposes of FMEA development, it is critical to identify aspects of the system design where a single event could cause the loss of more than one component leading to the system failing to perform its intended function. In conducting the FMEA, consideration should be given to external factors such as temperature, humidity and vibration which can lead to common causes in redundant systems. An example of common cause failure might be a common power supply provided for redundant displays. Common connections between systems create paths by which a fault in one system may affect another independent system. This is particularly true for redundant systems. Certain connection points are not only unavoidable but advantageous. The FMEA should consider the impacts of failure propagation and ensure adequate mitigation exists or is proposed.

Example: common cause failures. Simultaneous failure of two cooling water pumps due to power failure caused by damage to an electrical cableway, which contained power supply cables for the pumps, resulting in blackout conditions. Simultaneous failure of two computer networks due to software-related failure, resulting in the total loss of computer functionality. Loss of communications causes both master and slave drive to fail simultaneously because slave drive does not receive status of master drive and fails to automatically take over upon loss of master drive.

Refer to IEC 60812, Section 6.1 for further guidance on Common Cause Failures.

Unavailability of Redundancy (due to maintenance or other cause)

In redundant systems, the second system provides an extra level of safety. For redundant Classi-fication notations such as redundant propulsion or DPS-2 or DPS-3, the operational philosophy expected by Classification is that during normal operations both systems are available. If one of the redundant systems is down for maintenance, the normal operations should cease until both systems are functional. Therefore, analysing an FMEA scenario of failure of one system while the redundant counterpart is down for maintenance is not a requirement of Classification because such operational philosophy is not contemplated by Classification. However, the owner/stakeholder may wish to perform an FMEA to include a reliability focus. A typical example may be where a particular system has a history of unreliability with certain components, requiring some part of the system to be down for maintenance on a regular basis. In this case, it might be desirable for the owner/stakeholder to ask that the analysis be performed not merely including a single failure of the unreliable component, but with the assumption that one of the unreliable components will already be unavailable at any given time during operations due to known main-tenance issues. Such analysis, combined with criticality, may point out the more likely failures, the most critical and would allow for corrective actions to increase the system reliability.

Failure of Active and Passive Components

Certain Classification FMEA, such as those required for DP Systems, make use of a concept of passive and active components when deciding what types of failures will be included in the FMEA. The concept of passive and active equipment can be explained as follows:

- Active or rotating components in mechanical systems refer to machinery that moves and rotates during operation (e.g., pumps, compressors, generators, thrusters, remote controlled valves, etc.). For electrical/electronic systems, active equipment refers to those that require being powered in some way to make them work (e.g., integrated circuits, PLCs, switch-boards, etc.); and
- Passive or static components in mechanical systems refer to those having parts that normally do not move (e.g., pipes, tanks, vessels, shell-and-tube exchanger, manual valves, etc.). For electrical/electronic systems, passive components are those that do not require energy to make them work (e.g., electrical cables, resistors, capacitors, etc.).

Passive static components are, in general, considered to be of high reliability, whereas active components have a lower reliability. However, unless otherwise indicated (e.g., DPS-2 FMEA), Classification position is that even passive components can have a significant probability of fail-ure in mechanical systems (i.e., small diameter pipes, gaskets, flanged connections in the pipe, etc.) and its failure should be considered in the FMEA and demonstrated to be mitigated to an acceptable level.

External Events as Failure Modes

FMEA purists will contest that external events leading to equipment failure shall not be discussed in FMEA. Traditional FMEA are about assessing the impact of the failures that originate within the equipment. However, the coverage of credible external events is needed to fulfil the intent of many Classification-required FMEA. The system response to such external events such as fire and explosion in the vicinity, flooding scenarios, or adverse environmental conditions such as hurricanes and typhoons should be analysed. One way to include these items in the analysis is to include them as the "failure mode," and then proceed to complete the FMEA. A clear example of this is the FMEA for a DPS-3 system. The intent of a DPS-3 FMEA is to prove that the system is not only redundant, but that there exists physical separation between redundant systems such that an event external to the DP system (e.g., fire or flooding) will not compromise both systems and result in the loss of the DP functionality. In these cases, it makes sense to list the external event as a "failure mode" and analyse its impacts and existing control measures. The analysis of these external influences shall analyse if the:

- Equipment in question is designed to safely react to the external event.
- External events can produce equipment failure of interest, including damaging the existing risk control measures; and
- External events can cause multiple failures (common cause).

As an example of common-cause failures caused by external events, let us assume a fire near the diesel firefighting pump. Due to its location, the fire can damage nearby equipment which includes firefighting equipment, thus making the extinguishing of the fire difficult.

FMEA of Controls, Instrumentation and Safety Systems

There are multiple instances of in the Class Rules and Guides of requirements to conduct FMEA for instrumentation and safety systems. These FMEA are to consider the failure of the components within the instrumentation and safety systems, and its effects to verify that no hazardous consequences arise from their failure. However, it is important to note that analysing the failure of a safety system (safety ESD, alarm, etc.) will not result in a consequence unless coupled with a demand for its functionality. This demand for the safety system is caused by a problem in the system it is protecting or controlling. Therefore, failures of the equipment under control should also be considered as to ascertain the adequacy of the safety system to protect the equipment.

ANALYSE EFFECTS

Each failure mode is analysed in terms of possible consequences on operation, function, or system status. The failure mode under consideration may have a larger effect than just on the one element or function under study. So, in addition to the local effects, wider global effects are also considered. Particular attention should be paid to the impact a failure will have on the overall functionality of the system and how the system will react/behave after the failure is realised. The consequences of each identified failure affecting the item are captured in the analysis to provide a basis to evaluate any existing safeguards or to allow recommendation for corrective actions. Note that operational modes and interfaces become key factors to analyse the properly analyse global effects of failures. Chapter 2 discusses functional and physical interfaces in FMEA. When analysing effects of failure in one system, the effects on their interrelated systems must also be analysed to give an accurate picture of the effects beyond the local system. Equally important for adequate analysis of local and global effects is to analyse failures within the context of each relevant operational mode as discussed in Chapter 2. Figure 3.2, and its explanation illustrate the relationship between operations and the FMEA. The FMEA should identify the end effect of any failure in terms of the impact on safety and the operation, as in many cases, unsafe situations derive from negative impacts of failure on the operations.

Effects of Interest (Undesirable Events)

The end goal of an FMEA is to evaluate whether enough measures are in place to prevent the occurrence of a hazardous or otherwise undesired event due to a single failure. Examples of such events could be:

- Injury.
- Pollution.
- Loss of integrity; and
- Inability to perform a function (such as maintaining position in a DP system, or propulsion in a Redundant Propulsion system, or well control capability in a BOP system).

Failure can result in consequences with limited local impact to the system where the failure occurred or in wider impacts to the vessel or installation. A leak of flammable fluid from a pump can result in a loss of the pump, loss of the system and a fire that can spread and have effects beyond the immediate system of interest. All the potential consequences should be discussed, it is important to determine as part of the FMEA scoping what are the undesirable events or effects of interest. Before the FMEA, these effects of interest should be communicated to all participants to help focus on the analysis. During the FMEA, the possible realisation of these effects should be considered in every failure and clearly documented if the possibility exists. This approach minimises the possibility of FMEA that do not satisfy their intent because they do not clearly discuss or document if the effect of interest could be realised upon a particular failure.

It is a best practice for FMEA to consider the worst credible potential effects that could result from the failure. This includes considering what could happen if the risk controls that are supposed to detect, prevent or mitigate the failure do not work because they are themselves in an undetected failed state (hidden failures) or due to common-cause failures (common precursor led to both failures).

Refer to IEC 60812, Section 5.2.5 for further guidance on Failure Effects.

IDENTIFY FAILURE DETECTION METHODS

A given failure mode will manifest itself in a way which can be observed through system behaviour and through any number of indications. A necessary component of an FMEA is the identification of failure detection methods for each failure mode. Detection can occur by various means including visual or audible alarms, sensor limit warnings, sensor health and status checks, data comparison algorithms, operator observation, etc. By identifying detection methods, an assessment can also be made concurrently to determine if detection methodology is sufficient to accurately identify the failure, either by the system or the operator. If insufficient detection exists (refer to Chapter 3) or if the failure can be easily misdiagnosed leading to a larger issue, corrective actions should be made to add additional detection methods (e.g., additional sensors, operator check lists). Adequate time must be available to react in case operation action is required to reach a safe state or prevent escalation. If this is a concern, failure detection is not sufficient, and prevention of the specific failure mode is necessary.

IDENTIFY EXISTING RISK CONTROL METHODS

This is part of the analysis where we take note of all the existing protection mechanisms to prevent failure or to mitigate its consequences. In a newly built situation, only the risk control methods shown in the drawings should be assumed to be in place in the design. If a particular risk control method is not shown in the drawings, and the FMEA team deems it is needed, then corrective action should be taken so that it will be implemented. When analysing risk controls, it is important to consider their potential for hidden or latent failures. Risk controls tend to be safety equipment or equipment that performs a back-up function and is in a non-operational or standby mode. Without proper monitoring to detect a failure of the risk controls, such failure is only discovered when there is a demand for risk control due to another failure. Since the initial hidden failure is unknown until the second failure occurs, the two failures can be considered together as a single event, and thus analysed together in the FMEA. It is important to note that not every hidden failure will necessarily be assessed. The level to which hidden failures are assessed depends on the goals and intent of the analysis. Typically, the analysis will be limited to hidden failure/additional failure combinations that lead to an undesired event, but loss of either component on its own will not. Refer to IEC 60812, Section 5.2.7 for further guidance on "Failure Compensating Provisions".

CRITICALITY RANKING (FOR FMECA)

Refer to Chapter 2 of this book as well as IEC 60812, Section 5.2.8 and 5.2.9 for further guidance.

IDENTIFY CORRECTIVE ACTIONS

If the analysis indicates that the undesirable consequences can result from a single failure, corrective actions should be suggested to demonstrate compliance with the class design philosophy. Typical solutions suggested during FMEA to correct identified failures in a manner compliant with the design philosophy set forth in the Classification requirements are:

- Redundancy in design (this may be the only corrective action acceptable if the system functionality must continue after a single failure).
- Safe and controlled shutdown; and
- Actions to reduce the likelihood of failure.

Any repercussions or changes deriving from implementation of the FMEA corrective actions should also be noted and submitted to Classification such as modifications to maintenance procedures or schedules, updates to drawings, documentation, etc. Recommendations can be categorised into priority groups, for example, "Classification Requirement," "For Immediate Attention," "For Serious Consideration," and "For Future Improvement. Decisions." It is the responsibility of the entity engaged on the contract with Classification (i.e., shipyard, owner, etc.) to follow through on the corrective actions needed to comply with Classification requirements.

Chapter 4

FMEA Report and Class Review of the FMEA

FMEA REPORT

The FMEA report should contain sufficient system information for the reader to understand the stated failure modes, effects, existing risk control measures and related recommendations. In addition to a detailed narrative description of a system, failure modes and effects narratives should also be included to describe each of the relevant failure modes and effects for a given system.

FMEA REPORT STRUCTURE

An FMEA report is structured according to the scope of the study. At a high level, the parts of the FMEA report should include an executive summary; an introduction and background to the report; a description of the systems under review; a section with conclusions and corrective actions; a collection of appropriate FMEA Worksheets; and records of the reference data.

The executive summary should present the conclusions and corrective actions of the FMEA, pending as well as those corrective actions already implemented. The introduction should give all the background information needed so the report can later be understood and used by someone that did not participate in the study. It should include a statement of the purpose of the FMEA, when it was developed, who the participants were, assumptions, approach, etc. The description of the system(s) should include a narrative as well as a block diagram graphically identifying the scope of study. The block diagram is helpful to the team in preparation for the analysis as well as a valuable resource during the analysis and to anybody reviewing or consulting the final FMEA report. Each major system within the scope of the study is given its own section of the document which should contain enough descriptive information for the reader to understand overall layout, design and functionality, along with sufficient data (including functional schematics, drawings, etc.) to allow comprehension of the failure modes and effects. The report should contain descriptions of identified failure modes, causes and effects as they pertain to the subject of the study, a summary of any conclusions or recommendations, and any outstanding or unresolved action items.

The FMEA worksheets are typically used to record, in a tabulated format, failure modes, causes, effects, detection methods, safeguards and recommendations throughout the FMEA process. These can be included at the end of each system section or compiled in their entirety as an appendix. When FMEA proving trials are planned or required, there should be a trials programme report that is either a standalone report or incorporated as part of the FMEA report. The trials programme report is to provide test sheets for failure modes identified through the analysis. The test sheets will contain methods for testing, procedures to perform the tests, results from the tests, and any comments or recommendations. Results from trials may influence the content of the FMEA report and the FMEA should be updated accordingly. A typical FMEA report contains the information depicted in Table 4.1.

FMEA INTERNAL REVIEW PROCESS

The review process during FMEA development is somewhat iterative in nature. A provisional report with preliminary recommendations will be supplied to the owner/stakeholder for review. Any modifications, technical issues or areas of interest will be discussed among the FMEA team and stakeholders so that the analysis and recommendations are clearly defined and understood. If any modifications are necessary, the FMEA team will appropriately amend the document and deliver a definitive version for acceptance.

CLASSIFICATION REVIEW OF THE FMEA

The entity that has signed the contract with Classification is ultimately responsible to see that an FMEA report that satisfies the intent of the FMEA requirement is submitted to Classification. In the majority of cases the FMEA will be submitted directly by the vendors contracted and Classification will use the FMEA conclusions as evidence that the design is in compliance with the Classification philosophy requirements (i.e., redundant systems, no single failure leading to unsafe situation, etc.). The FMEA should refer to the version of the design submitted for Classification approval. Any revisions to drawings or documents referenced in the FMEA report may affect the findings of the FMEA and require updating of the FMEA as necessary and submitted for review. If the FMEA was conducted with additional goals over and beyond meeting Classification requirements, such as optimising the design, identifying situations critical to operations, or developing a reliability-centred maintenance plan, it would facilitate the review if the Classification items were somehow highlighted.

Table 4.1 Sample FMEA Report Structure

Sections	Subsections	Description
Executive Summary	Intent of the FMEA	This summary provides a global assessment of priority issues and recommendations, if any. It can be presented in a tabulated format showing the overall number of failure modes, causes, criticality ranking of the failures, if carried out. If the FMEA identified any issues that need attention to reduce risk and comply with Classification requirements, a list of the plan actions should be provided in order to make the system comply with the Classification requirements.
	Summary of FMEA Conclusions	Additionally, for large and complex systems, it helps to provide a summary of conclusions for each major subsystem or equipment.
Introduction and Background	General	General description of the scope and purpose of the document in an introduction of the FMECA. This is a concise, aggregated description of the purpose of the FMECA, and should include the date it was conducted, the intended audience of the FMECA, as well as its contents.
	Application of Class Rules	Refer to the applicable Rules, desired Classification notation and specific requirement for which the FMEA is developed.
	Scope and Intent of the FMEA	This is a description of the scope and intent of the FMEA. Should include the intended audience, as well as its contents.
	FMEA Version and Date	Identify if prior FMEA have been conducted, what are the changes and the version for the current document, as well as dates when prior and current FMEA were conducted.
	Design Changes	If the system in question has experienced design modifications since the previous version of FMEA or there have been modifications to supporting systems or other systems which may impact system under analysis, describe those changes.

Table 4.1 Sample FMEA Report Structure (continued)

Sections	Subsections	Description
	Participants and Reviewers	List the names and roles of the FMEA participants, facilitators, reviewers and editors. This provides a source of informed personnel and/or accountability. For multiple revisions of an FMEA, provide a log listing the names and roles as well as which version they reviewed or edited.
	Key Design Concepts	Describe the key design features for the system redundancy, including the redundant features of the supporting utilities. This is only applicable for systems required to have redundancy and for which the FMEA is performed in order to determine the existence and adequacy of the redundancy.
	Main Equipment (Physical Boundaries)	Identify the pieces of equipment and machinery that were analysed in the FMEA. List the higher level system/subsystems and individual equipment, as applicable.
	Modes of Operations (Operational Boundaries)	Description of all modes of operations considered. This should include both the high level operations of the vessel/installation, critical activity modes or missions, as well as the modes of operation of particular systems analysed.
	Vessel Overview and Specifications	Provided for quick definition reference.
	FMEA Approach	An introduction and explanation of the process as a way to inform the reader how to interpret the results of the FMEA. If risk assessment and risk matrix were utilised, provide the risk matrix and describe how it was used.
	Ground Rules and Assumptions	Ground rules could be type of failures, ultimate consequences of concern, excluded items, etc. Needed for transparency of the analysis so that all users and reviewers of the FMEA understand the basic ground rules and assumptions.
	Glossary	Abbreviations and definitions provided for quick reference
Description of Systems		Failure modes and related recommendations, per system. For example, for a DP FMEA, these sections would include:
	Subsystem 1	• Power Generation
	Subsystem 2	• Power Management

Table 4.1 Sample FMEA Report Structure (continued)

Sections	*Subsections*	*Description*
	Subsystem 3	• Propulsion and Thrusters
	Subsystem 4	• DP Control
	Subsystem 5	• Other
Conclusions/Corrective Actions		An overall assessment of the findings of the FMEA. List of corrective actions that originated from the FMEA. If none, are all the failure modes analysed in compliance with the Classification design philosophy?
References		List of drawings with revision numbers, manuals, etc., used to develop the FMEA.
Appendices	FMEA Worksheets	Completed FMEA worksheets.
	Table of Abbreviations	Table of abbreviations used in FMECA is provided for quick definition reference.
Verification Plan	Selected tests, procedures, and expected results.	A verification plan to validate the conclusions of the FMEA is not required for every FMEA. This may be a separate stand-alone document. Refer to Chapter 5 for details.

PITFALLS AND COMMON PROBLEMS IN CLASSIFICATION SUBMITTED FMEA

This section highlights typical problems encountered in FMEA submitted to Classification. There are issues that may be specific or more prevalent among certain systems, but the reader can use this list as a check sheet to avoid making these common pitfalls. Guidance on how to avoid these pitfalls is given elsewhere in this report and the specific chapter also referred to in the information in parenthesis.

1. Scope
   * Parts of the system omitted in the analysis (refer to Chapter 2, Chapter 3, and specific requirements in Chapter 7).
   * Critical operations omitted the analysis (refer to Chapter 2, Chapter 3, and specific requirements in Chapter 7).
2. Failures
   * Incomplete failure list (refer to Chapter 3 and specific requirements in Chapter 7).
   * No consideration of common cause failures (refer to Chapter 3).
   * No consideration of hidden failures (refer to Chapter 3).
3. Effects
   * Global end effects not addressed (refer to Chapter 3).
4. Controls
   * No consideration of hidden failures on existing controls (refer to Chapter 3).
5. Corrective Actions
   * Failure of or delayed follow through of corrective actions (refer to Chapter 3).
6. Overall
   * Insufficient descriptions in the worksheets to understand the failure scenarios (refer to Chapter 3).
   * Insufficient information in FMEA Report (refer to Chapter 4 and the specific requirements in Chapter 7).
   * FMEA not matching the latest design or off the shelf FMEA (refer to Chapter 2).
   * Submissions too late (refer to the specific requirements in Chapter 2).

FMEA AND SUPPORTING DOCUMENTATION SUBMISSION

Items required to be submitted to Class to support the FMEA are listed in the individual Classification Rule requirements. However, as a general rule, the following documentation enhances understanding of the system and the FMEA, and it should be submitted to Classification as applicable:

* FMEA Worksheets (Mandatory), indicating revision number, date.
* FMEA system Boundary Description.
* System Design Specifications.
* Functional or Reliability Block Diagram(s) showing interactions of all systems.
* Control system details.
* Cause and effect matrix (useful in particular for items such as control, PLC, and safety systems).
* One-line diagrams.
* Bill of materials.
* Operating procedures manual.

- Detailed narrative of system functional description.
- Piping and instrumentation diagram (P&ID).
- Basic schematics or equipment drawings.

- General arrangement.
- Process flow diagrams (PFD).
- Emergency procedures.
- Maintenance, Inspection, Testing (MIT) procedures.
- FMEA validation plan and procedure, if required by Classification.

As most of the items in the list above have been discussed elsewhere or are self-explanatory, they will not be expanded upon here. The cause-and-effect matrix and the operations manual have not been discussed before and are clarified in the next chapter.

It must be emphasised that the FMEA should refer to the version of the design submitted for Classification approval. Any revisions to drawings, documents, functions, or operations referenced in the FMEA report may affect the findings of the FMEA and require updating of the FMEA as necessary and submitted for review.

CAUSE AND EFFECTS MATRIX

The cause-and-effects matrix was originally derived from Safety Analysis Function Evaluation (SAFE) Charts in API RP 14C for offshore facilities for documenting safety requirements. It is a straightforward way for those familiar with the equipment and operations to understand the logic being implemented in the safety system. A cause and effects matrix identifies the potential causes (or deviations), listed in rows down the left side of the matrix. The system responses and their resulting effects are listed in columns across the top. The intersection cell in the matrix defines the relationship between the cause and the effect and aids in understanding the system safety control logic.

Fig 4.1 Sample Cause and Effect Matrix

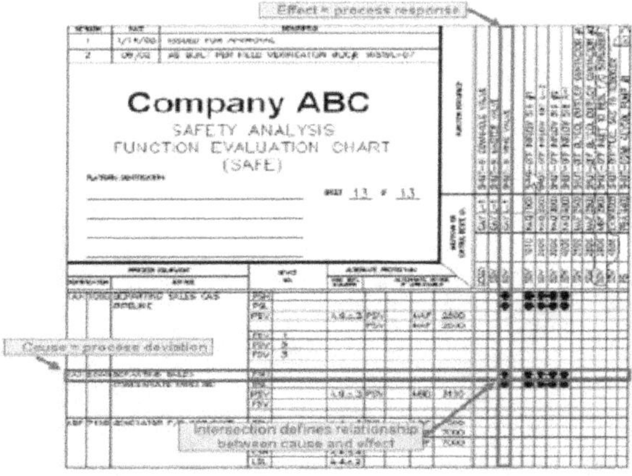

Diagram courtesy of CFSE, Certified Functional Safety Expert

OPERATIONAL MANUALS (OPERATIONS, INSPECTION AND MAINTENANCE, AND EMERGENCY)

The requirement to submit operational and maintenance procedures to Class serves several purposes during the classification process:

1. To demonstrate compliance with design and manufacturer's requirements.
2. As an aid to understanding how the system will be operated, and what constraints, if any, are put in the operation of the system (e.g., for redundant systems, are there any operational limitations defined for the case when one system is unavailable?)
3. To confirm that FMEA findings have been incorporated into the equipment operating phases, as needed.

Recommendations from an FMEA performed early in the design process can usually be implemented to eliminate or reduce the adverse effects of a failure. However, when it is not possible to do so, operational and maintenance procedures and plans can be developed and/or updated to mitigate failures and their effects, in addition to providing a means to communicate actions to be taken when a failure does occur.

Chapter 5

FMEA Verification Programme

PURPOSE

The FMEA is a tabletop study, and it alone may be insufficient to provide a satisfactory level of assurance. Thus, the FMEA process may also incorporate a trials programme to verify that the systems will perform as predicted in the analysis. Trials prove the FMEA in a structured manner. The specific tests included in a trials plan are designed to verify conclusions reached in the FMEA study. For the purposes of this book, the terms FMEA testing, FMEA proving trials, FMEA validation and FMEA verification programme are equivalent. They refer to trials and testing necessary to prove the conclusions of the FMEA or to establish conclusively the effects of failure modes that the FMEA desktop exercise had a high degree of uncertainty about. The standard term used in this book is FMEA verification programme. A verification programme is not always required as part of an FMEA study. The inclusion of a verification programme will be determined by Classification requirements for the particular system or by the owner/stakeholders goals. FMEA verification programmes for Classification have traditionally been limited to DP systems, certain drilling equipment for the CDS notation, and to a lesser extent for electronically controlled diesel engines. However, many stakeholders choose to perform an FMEA verification programme for other types of systems simply to validate the conclusions of the FMEA, even though such programme may not be required by Classification.

SCOPE OF THE FMEA VERIFICATION PROGRAMME

A question to be resolved early in the planning of the FMEA verification programme is what failures are to be verified. The scope of the testing will be established by the FMEA team to comply with the Classification requirements, as well as other goals specified by the owner/stakeholder. In general terms, FMEA results that should be included in the verification programme are as follows:

1. Those areas of a mechanical or control system which has mitigating barrier(s) to prevent the occurrence of a hazardous situation. The verification programme is to validate that upon the specified failure, a minimum of one mitigating barrier performs as intended in order to prevent the occurrence of a hazardous situation. Examples of such mitigating barriers are hydraulic load holding valves, alarms, sensors, etc.
2. Those results for which there is reasonable uncertainty or disagreement of the FMEA assumptions.

During the FMEA, uncertain assessments results are to be identified and discussed with the designer at time of Design Review for resolution. If a resolution is not achieved, these items should be included in the verification programme. Examples of areas where inadequate data may be available to perform definitive analysis, thus should be part of the verification programme include the behaviour of interlocks that may inhibit operation of essential systems. It is the best practice to identify, as part of the FMEA scoping and during the FMEA, the verification necessary to prove the FMEA conclusions. Other items traditionally evaluated during FMEA verifications include system wiring checks; control software functionality and response to failures; confirmation of system's ability to operate with failures, in accordance with design intent; confirmation of system response to common system failures; and for redundant systems, confirmation of continued functionality after failure of a redundant system.

When the Classification requirement necessitates that FMEA trials be conducted, the trial plans shall be submitted to Class with enough lead time to allow for Class review and input both by Class Engineering and Survey. Test procedures are to be submitted for approval and retained aboard the vessel. Test techniques must not simulate monitored system conditions by maladjustment, artificial signals, improper wiring, tampering, or revision of the system unless the test damages equipment or endanger personnel. As the Classification Surveyor prepares for witnessing the verification programme, he or she will review and compare the FMEA, the verification plan and the current vessel drawings. The Classification Surveyor has the discretion to modify the verification plan to change or increase testing if he or she realises that the FMEA does not accurately reflect the vessel or the vessel systems under trial (e.g., changes made to the vessel after the FMEA was developed, inaccuracies or gaps in existing FMEA).

The verification plan will be submitted to the owner/stakeholder for review and input prior to testing. It may be a conflict of interest when the entity developing the system is also in charge of finding its faults and testing for them, for example, a new-built situation where verification plans are typically developed by the shipyard. It is advantageous that other stakeholders, the owner in particular, have an input in the development of these plans. The verification programme can be conducted in a number of ways depending on the tests to be performed or functions to be verified. Tests that can be performed dockside or at the vendor facility will be identified so they can be verified independent of sea trials, if desired. If equipment FMEA tests are performed at the vendor facility, there may be a particular item regarding integration with the complete vessel or facility that can only be evaluated after integration.

Alternate Testing Methods

Certain verification tests may not be feasibly conducted, such as when a system cannot be evaluated without causing damage to hardware or creating a situation that would be imminently hazardous to personnel. For this case, an alternative test method should be proposed together with an explanation of why it is an equivalent test. If failure cannot be reproduced, at the very least, the safeguards to protect in case of such failure shall be evaluated to verify their existence, specifications, functionality, maintainability, and methods by which a safeguard failure will be made evident. Non-traditional tests, such as hardware-in-the loop (HIL) testing and self-diagnosis test, may be accepted on a case-by-case basis.

Fig 5.1 FMEA Trial Test Sheet Example

EQUIPMENT: REFERENCE   SYSTEM
Test #   HiPAP   Test   FMEA Reference #:
 Method :
- On DP.
- All thrusters on line.
- HiPAP   and DGPS online

1. Interrogate transponder and select as reference on DP.
2. Lift transponder used for DP without   deselecting.
    a)   Select transponder to DP when suspended to test   voting.
    b)   Make small move.
    c)   Stabilize .
3. Fail power to   HiPAP transceiver unit. DPUPS5   Vacon Room.

Results expected:
1. DP uses   HiPAP   and it agrees other references.
2. Transponder rejected when moved. Transponder rejected when suspended.
3. Alarm . HiPAP rejected on power failure .

Results found:

Comments:

Witnessed by: Date:

VERIFICATION PROGRAMME TEST SHEETS

The verification plan, test sheets and associated procedures are created by the FMEA team and reviewed by the owner and Classification prior to being performed. Test sheets are developed based on the established scope and identified failure modes from the FMEA report. Each test sheet usually represents a single set of tests for a given component or function. Each test sheet must be in a step-by-step or check-off list format and should include:

- Description of the hardware/function being evaluated.
- Purpose of the test, linking it to the FMEA and specific failure of concern.
- Test methodology, including apparatus necessary to perform the test.
- Procedures to perform the test (including equipment status, safety precautions, safety controls and alarms set points).
- Expected results (these are subject to change for first-time tested equipment based on results of initial testing and agreement from all parties involved).
- Evaluate points and success/acceptance criteria.
- Space to describe actual results and any comments an example test sheet is provided in Figure 5.1.

PERFORMING FMEA VERIFICATION PROGRAMME

The stakeholders perform the verification tests, usually with the oversight of representatives from the FMEA team and attended by a Classification surveyor, if it is a Classification requirement. Test sheets are provided to the owners/stakeholders in advance of the verification tests to allow time for review and comment. This also provides the stakeholders with an opportunity to understand the tests that will be performed, and that proper coordination has occurred on board to allow testing to go smoothly.

RESULTS AND RECOMMENDATIONS

Overall results of the FMEA verification programme are discussed on board and any potential issues presented so that the owner/stakeholder will have an early opportunity to address items of importance prior to a preliminary report being delivered. A short one or two-page summary of issues and recommendations is provided by the FMEA trial team prior to their departure from the site. Recommendations may include required system changes or mitigation strategies to meet Classification or may be simple items that would improve system operability but are not required to satisfy any requirements. If the system does not work as expected during a trial even though the FMEA and the trial plan seem to accurately reflect the as-built condition, a need for repair is normally identified to correct the situation. If the system does not behave as expected in the FMEA because the analysis does not accurately reflect the as-built condition or because of an overlooked engineering issue, the Classification Surveyor writes an outstanding requirement in his or her report and requires the modification to be tested once in compliance with the Rules.

FMEA VERIFICATION PROGRAMME REPORT

The complete FMEA verification programme report with test results and recommendations should be developed. If being developed by a third-party, this complete document should be reviewed and accepted by owner/ stakeholders and submitted to Classification. There will be a dialogue between parties throughout the process so that recommendations are appropriate, corrective actions are properly implemented, and requirements are met to the satisfaction of Classification and, if involved, the FMEA team. Close-out items are tracked within the document so that there is a lone source of information and tracking for all parties involved to reference. Responsibility for closing-out items as well as additional revalidation tests shall be indicated. A typical verification programme report consists of a set of tests organised by system and comprised of numerous individual test sheets. Once the verification tests are performed, the complete FMEA report should capture the test data verifying the conclusions of the FMEA. It is recommended that the List of Alarms (i.e., signal list, inputs/outputs list) be provided as an appendix to the FMEA verification programme report so that alarm titles, set points, and time delays can be verified during testing. In many cases, during a test, unforeseen alarms may occur, and it is important for the Surveyor to be able to verify if the anticipated alarms specified in the test procedure actually occur. In some cases, failure may have unintended consequences that were not expected. A sample FMEA verification programme report structure (using a DP example) is shown in Table 5.1.

UNITED STATES COAST GUARD DESIGN VERIFICATION TEST PROCEDURE

For vessels under the United States flag which must undergo the U.S. Coast Guard (USCG) regulatory body review of a vital automation system, this phase of the approval process is known as the Design Verification Test Procedure (DVTP). A Design Verification test is to be performed once, immediately after the installation of the automated equipment or before issuance of the initial Certificate of Inspection (and thereafter whenever major changes are made to the system or its software), to verify that automated systems are designed, constructed and operate in accordance with the applicable Class rules and requirements of this supplement. The purpose of design verification testing is to verify the conclusions of the qualitative failure analysis (QFA). The DVTP is therefore an extension of the QFA and the two may be combined into one document. DVTP should demonstrate that all system failures are alarmed and that all switchovers from a primary system component to a back-up component are also alarmed. Design Verification and Periodic Safety test procedures are to be submitted for approval and retained aboard the vessel. Test procedure documents must be in a step-by-step or check off list format. Each test instruction must specify equipment status, apparatus necessary to perform the tests, safety precautions, safety control and alarm set points, the procedure to the followed, and the expected test result. Test techniques must not simulate monitored system conditions by maladjustment, artificial signals, improper wiring, tampering, or revision of the system unless the test damages equipment or endanger personnel. Where a test meeting the restrictions on test techniques will damage equipment or endanger personnel, an alternative test method shall be proposed together with an explanation of why it is an equivalent test. DVTP is a detailed test procedure to verify each failure mode identified in the QFA. Each test should include:

1. Safety precautions.
2. Equipment status prior to testing.
3. Equipment required to perform the test.
4. Control or alarm set-points.
5. Evaluate procedure to be followed.
6. Expected results; and
7. Space for the cognisant Officer in Charge, Marine Inspection (OCMI) or Authorised Classification Society (ACS) Surveyor to record results during testing.

Table 5.1 Sample FMEA Verification Programme Report Structure (for a DP FMEA)

Main Section	*Subsections*
Executive Summary	
General Background	• Introduction. • Scope of Work. • Conduct of Trials. • Personnel. • System Limitations. • Reference Documentation. • Vessel Overview and Specification.
Trials Findings and Conclusions	
Recommendations	
Equipment Status and Records Verification	

Table 5.1 Sample FMEA Verification Programme Report Structure (for a DP FMEA)
(continued)

Main Section	*Subsections*
Closeout Tabulations	
Trials Programme	• Test Sheets for Each System • Power Generation • Power Management • Propulsion and Thrusters • DP Control
Appendices	• Supplemental/Supporting Data from Trials • Action Item Close-out Communication/Status • Alarm List, Signal List, Input/Output List with alarm titles, set-points, and time delays.
Trials Report Addendum	

Chapter 6

FMEA Lifecycle Management

**BEST PRACTICES FOR TREATING THE FMEA AS A
LIVING DOCUMENT**

As general rule, Class does not address the management of the FMEA after granting of the Classification or special notation. However, Class should be notified of any changes such as design, function, operations that impact the basis of the Classification requirements and it will be up to the discretion of the Class reviewer whether to require a revised FMEA.

In certain notations, such as CDS and DP, it is not discretionary, but a requisite that any changes to the design, function or operations of the systems that may affect the findings of the FMEA requires that the FMEA be updated as necessary and submitted for review. Such requirements are listed in Chapter 7 as applicable.

The information contained in this chapter describes the best industry practices. Even though Classification may not address FMEA management specifically, the asset owner should pose the following questions to decide what is desirable and feasible for their particular situation:

- Is the FMEA to be kept on board as reference? This may not be possible for vendor-performed FMEA, as the information within the FMEA may be vendor-proprietary. However, if the FMEA are developed with this additional goal in mind, the information that makes them proprietary may in some cases be omitted without compromising the quality of the FMEA.
- Shall the FMEA be updated regularly or after a change? If so what kind of change should trigger an FMEA revision?
- If the FMEA was originally performed by the vendor, who would perform any updates required as result of a change? Ideally, the vendor would be involved in the updating FMEA.

As per industry standard practice, certain FMEA, in particular the FMEA and verification programme (trials) reports for DP systems, are living documents that are intended to be maintained through the life of the system to which they pertain. They should be kept up to date on board for use by staff as required. The reports should be updated to reflect the latest information on a given subject including any system upgrades, modifications in configuration or changes to operational setup.

BEST PRACTICES FOR TREATING THE FMEA AS AN OPERATIONS RESOURCE DOCUMENT

An important use for the FMEA developed during the design phase of a system is to be a resource for operations and training for the crew. The FMEA can be a reference document to improve operator's understanding of the risks and corrective actions in place shall a particular failure occur. For example, the FMEA can be used to find out, upon detecting a particular failure, information such as:

- The predicted consequences and impact on other systems.
- What corrective action shall be taken?
- The impact on the redundant features of a system if an item of equipment is taken out of service.

The findings of the FMEA are incorporated into the operations, maintenance, emergency, and training manuals. In the case of DP FMEA, it is a best practice to incorporate the FMEA findings within the Well/Activity Specific Operating Guidelines (WSOG/ASOG) for mobile offshore drilling units (MODU) and offshore support vessels, respectively.

BEST PRACTICES FOR USING THE FMEA IN ASSET LIFECYCLE MANAGEMENT

Ideally, proprietary issues can be worked out with the vendor and the owner of the vessel or installation, so the FMEA conducted for the systems can be shared with the operating company that is responsible for the safe operation of the vessel or offshore installation. In certain cases, such as FMEA for DP, the full responsibility and ownership of the FMEA should be with the operating company. The DP FMEA is kept onboard and used as a reference, but the ownership of the FMEA typically resides within the management team ashore who is the responsible point for changes. It is not an individual's responsibility as such, though it is common that the vessel superintendent or the asset manager in the shoreside management office be designated the focal point and should have a thorough understanding of the FMEA management process. Key personnel onboard have the responsibility to make the shore management team aware of any deficiencies or inaccuracies in the FMEA as they themselves become aware of them. The management team is responsible for ensuring that any such deficiencies or inaccuracies are corrected in a timely manner. The FMEA should be identified as a controlled document embedded within the quality management system of the owner. Any changes to the FMEA content will be identified through the audit trial. It is essential that an FMEA change control management procedure is in place as part of the company quality or safety management system under the ISM Code. Adopting this procedure provides a formal process to track every change in the vessel or installation systems and to capture, record and feedback decisions to the vessel for implementation of corrective actions.

The change control management procedure must include a facility to feed back into the FMEA any failures in operation, including those not resulting in an incident.

Fig 6.1 FMEA Lifecycle Management

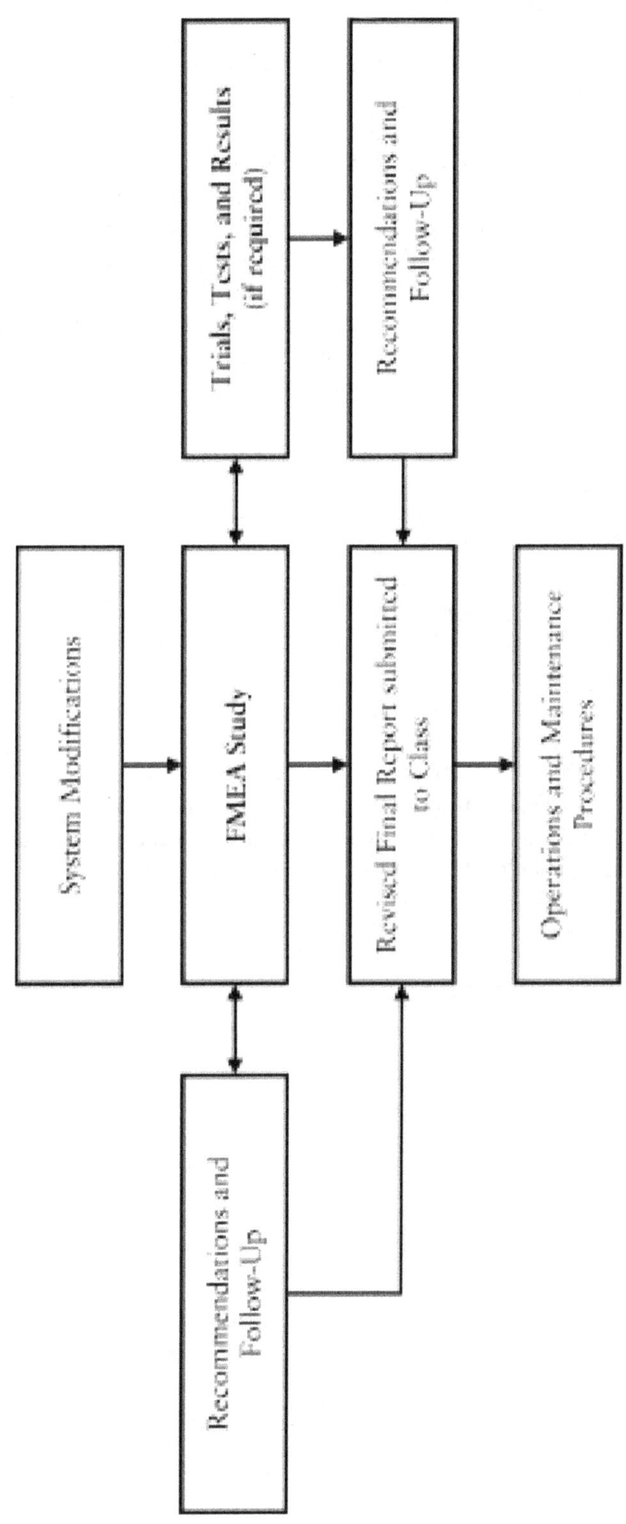

CHANGES TO THE CLASSIFICATION SYSTEM AND FMEA REVISIONS AND SUBMISSIONS

Any upgrades, modifications or changes that affect the FMEA must be evaluated to determine what action will be taken so that the documentation is coordinated with the latest system configuration. This may entail simply modifying an existing FMEA study and issuing a revised report or producing a completely new FMEA and report along with a new or modified verification programme. Modifications that pertain to a Classification requirement should trigger the resubmission of a revised FMEA report to Classification. It is a best practice that the FMEA report be periodically reviewed for accuracy as changes may creep in that are not obvious. It is a Classification requirement for certain systems, such as for DP systems, that a review of the FMEA occurs every five years, even if no apparent changes to the subject have occurred over that time. The idea is to catch hidden or unreported modifications or changes that have slowly and almost imperceptibly crept into the system and operations.

FMEA AND MANAGEMENT OF CHANGE

The FMEA is a tool to aid in the verification of vessel safety and compliance with certain Classification design requirements. Significant changes to the classed system may require updating of the FMEA to assist in verification that the changes did not compromise vessel safety or hinder compliance with Classification requirements. The FMEA should be governed by the company's management of change programmes. If no such programme exists, it is suggested that the updating of the FMEA be governed by the company's safety management system and a management of change form specific to the FMEA process be developed. An FMEA Management of Change form will include, as a minimum, the entries listed in Table 6.1 below.

Chapter 7 provides guidance on when the FMEA should be updated. As a general rule, Class needs to be notified of any changes made that impact the basis of the Classification requirements. It will be up to the discretion of the Class reviewer whether to require a revised FMEA.

Table 6.1 Suggested Entries in Management of Change Form for FMEA

General Descriptors	• Date • Name of Vessel or facility • FMEA Reference No • Systems Affected
Background	• Reason for change (i.e., incident, accident, unavailability of in-kind replacement, etc.) • Description of the change
Effects of Change on FMEA	• Does it affect the FMEA? • Has the FMEA been updated accordingly? • How was the FMEA modified?
Recommendations	• What recommendations came out of the FMEA update? • How were they implemented? • Which relevant personnel need to know about actions arising from the FMEA? • How are they communicated to relevant personnel?
Classification	• Do the change and the FMEA recommendations affect a Classification requirement? • If so, resubmit the revised FMEA and appropriate documents to Classification?
FMEA Verification/ Trial	• Test required? If so, what kind of tests? Have they been carried out? • Will test need to be witnessed by Classification?
Other documents	• Do the change and/or FMEA recommendations affect the Operations Manual? • Emergency operations manual? • Maintenance manual? • Drawings? • Training? • If the change does affect any of these documents, have they been changed accordingly?
Training and Communication	• Has the change and implemented FMEA recommendations been communicated? • Is training required? • Has personnel been trained?

Table 6.1 Suggested Entries in Management of Change Form for FMEA

Fleet Applicability	• Does this change and/or FMEA recommendation also apply to other vessels or facilities within the organisation? • If so, what action has been taken?
Circulation List	• Typical positions to be notified include Master or Offshore Installation Manager for an offshore installation, officers, Chief Engineer, maintenance, electrical, drilling crew, shore-based personnel, etc.
Signatures	• Appropriate signatures from vessel or installation management and/or supervisors.

Chapter 7

System-Specific FMEA Requirements

GUIDANCE FOR SYSTEM-SPECIFIC FMEA REQUIREMENTS

The use of risk studies in the industries served by Classification is becoming increasingly prevalent. The general elements of the FMEA process were discussed in detail in Chapters 1 through 6. Chapter 7 provides the detail and clarification of select FMEA requirements that appear in the Class Rules. Table 7.1 lists the different systems in Classification for which a risk study (i.e., FMEA or alike) is required and the subsection index to facilitate finding specific guidance. Faced with a particular Classification FMEA requirement, the user may choose to go directly to the relevant subsection as indexed in Table 7.1. The tables in this chapter can be used to clarify a certain requirement, much in the manner as one would use the dictionary to clarify a word. The guidelines address the following aspects for each individual FMEA Classification requirement:

- Purpose.
- Undesired events (e.g., consequences of interest).
- Systems or subsystems (physical scope).
- Modes of operation (operational scope).
- Typical failures.
- Timeline and team.
- Verification programme.
- Supporting documents; and
- Lifecycle management.

The explanation for what is covered in each of these bullet entries can be found in Table 7.2.

Table 7.1 Index of FMEA Requirements in Rules and Guides

Class Vessel Rules; Offshore Support Vessels (OSV); Under 90 Metres; Mobile Offshore Drilling Units (MODU); Mobile Offshore Units (MOU); Offshore Facilities; High Speed Craft (HSC); High Speed Naval Craft (HSNC); Gas Fuelled Ships; Propulsion Systems for LNG Carriers; Lifting Appliances

1	Automation	
	General Automation	
	Computer-Based Systems	
	Wireless Data Communications for Vessel Services	
	Integrated Controls	
2	Electronically-controlled Diesel Engine	
3	Remote Control Propulsion	
	Automated Centralised Control (ACC)	
	Automated Centralised Control Unmanned (ACCU)	
	Automated Bridge Centralised Control Unmanned (ABCU)	
4	Gas Turbine Safety Systems	
5	Redundant Propulsion and Steering	
6	Single Pod Propulsion	
Dynamic Positioning Systems (DP)		
7	Dynamic Positioning (DP) Systems	
Integrated Software Quality Management (ISQM)		
8	Software	
Mobile Offshore Drilling Units (MODU)		
9	Jacking and associated Systems	
Offshore Support Vessels		
10	Subsea Heavy Lifting	
Certification of Drilling Systems		
11	Drilling Systems/Subsystem/Equipment	
12	Integrated Drilling Plant (HAZID)	
Propulsion Systems for LNG Carriers		
13	Dual Fuel Diesel Engine	
Gas Fuelled Ships		
14	Reliquefication, Dual Fuel Engine and Fuel Gas Supply	
Lifting Appliances		
15	Motion Compensation and Rope Tensioning Systems for Cranes	

Table 7.2 Structure of the Guidance for Each FMEA Requirement

System	*Name of system for which the FMEA is required*
Rule/Guide:	Rule or Guide stating the FMEA requirement for the system above.
Requirement	This block is a textual copy of the Rule or Guide indicating the FMEA.
Purpose of FMEA	The purpose of an FMEA is to demonstrate compliance with the design philosophy for failure situations. The failure design philosophy will be stated as "a single failure shall not lead to a specified undesired event". The specified undesired event is typically one of the three listed in the section below. As part of the FMEA process, corrective action measures should be proposed to correct situations of non-compliance with the failure design philosophy. Identified non-compliances with the design philosophy should have been resolved or otherwise addressed by the time of submission of the FMEA.
Undesired Events	This section specifies the system and/or global consequences that could occur after a failure and that the design must address and avoid. These undesired consequences of interest or events fall within these following broad categories: 1. Loss of system or equipment function or degraded beyond an acceptable level 2. Loss of equipment control 3. Any unsafe situation with potential to harm individual, environment or equipment. Generally, Classification requirements cover systems whose functionality and control is critical to the safety of the vessel and personnel (i.e., propulsion in a ship, well control equipment on a MODU). For many of these Classification systems the loss of equipment or system function and loss of equipment control can ultimately result in an unsafe situation. Therefore, Classification design philosophy requirements for those systems are their continued functionality.
Systems or Sub-systems	This section lists systems and subsystems whose failures are required to be addressed in the FMEA to determine their compliance with Classification' design philosophy. The list is not exhaustive and is the responsibility of the entity contracting Classification to determine and analyse all the classed systems whose failure can result in the undesired event specified above.

Table 7.2 Structure of the Guidance for Each FMEA Requirement

System	*Name of system for which the FMEA is required*
Modes of Operation	A system normally has multiple modes of operation and each mode can present distinct failure scenarios. In particular, failure modes can be operation-specific and the resultant consequences of the failure can vary greatly depending on the mode of operation. The FMEA is to define the global operations at installation or vessel level (i.e., drilling operation, or pipe-laying), and local operations of the system/equipment which is part of the scope of the FMEA. The FMEA must analyse all the modes of operations (global and local) to identify failures, hazards and consequences of concern. This section will present a list of modes of operations that are to be explicitly considered on the FMEA. The list is not exhaustive and it is the responsibility of the entity contracting Classification to determine and analyse all the modes of operations during which a failure can result in the undesired event specified above.
Typical Failures	This section illustrates the types of failures that are expected to be analysed in the FMEA. The list is not comprehensive; does not address all possible failures and may include some that are not relevant for the design. All foreseeable failures are to be considered in the FMEA, even if not listed in this section.
Timeline/Team	This section suggests the optimal time in the system life to conduct the FMEA. It also suggests who, of all the parties involved in the Classification process, should have the main responsibility for the FMEA and which type of subject matter experts should participate in the analysis.
Verification Programme	If the system requires testing of FMEA results, this section gives a description of the expectations and responsibilities associated with testing the equipment, as well as the anticipated response to identified failures.
Supporting Documents	This section provides a list of the documents that aid in the review and understanding of the FMEA.
Lifecycle Management	This section indicates how the FMEA is to be used and updated during the operational life of the asset, as well as any requirement for resubmission to Class in case of changes that may impact the original basis upon which Classification was granted.
Additional Notes	Comments and notes not fitting in the other categories.

AUTOMATION (GENERAL CONTROL, SAFETY-RELATED FUNCTIONS OF COMPUTER-BASED SYSTEMS, WIRELESS DATA COMMUNICATION, INTEGRATED AUTOMATION SYSTEMS)

Class Vessel Rules; Offshore Support Vessels (OSV); Under 90 Metres; Mobile Offshore Drilling Units (MODU); Mobile Offshore Units (MOU); Offshore Facilities; High Speed Craft (HSC); High Speed Naval Craft (HSNC); Gas Fuelled Ships; Propulsion Systems for LNG Carriers; Lifting Appliances.

General Automation

Failure modes and effects analysis (FMEA) may be conducted during system design to investigate if any single failure in control systems would lead to undesirable consequences such as loss of propulsion, loss of propulsion control, etc. The analysis may be qualitative or quantitative.

Computer-Based Systems

FMEA is to be used to determine that any component failure will not result in the complete loss of control, the unsafe shutdown of the process or equipment, or other undesirable consequences.

Wireless Data Communications

A suitable risk analysis (such as Failure Modes and Effects Analysis (FMEA)) is to be performed which demonstrates that an interruption or failure in the wireless data `communication will not lead to a hazardous situation. *Note: Consideration is to be given to the possibility of corrupted data and intermittent failures with comparatively long recovery times between interruptions.*

Integrated Automation Systems

Where the integration involves control functions for essential services or safety functions, including fire, passenger, crew, and ship safety, an FMEA is to be conducted. The FMEA is to demonstrate that the integrated system will be 'fail-safe,' and that essential services in operation will not be lost or degraded.

Purpose of FMEA

Automation requirements, including the requirements to conduct an FMEA, apply to electrical, hydraulic, electronic, computer-based systems and equipment for control, monitoring, alarm, and safety on board vessels. Automation systems have Classification requirements to develop FMEA for:

- General automation for essential services.
- Safety-related functions of computer-based systems.
- Systems utilising wireless data communications.
- Integrated automation systems involving control functions for safety and/or essential services.

The purpose is to demonstrate that any single failure will not lead to an undesirable event such as:

- Lost or degraded function beyond acceptable performance criteria.
- Hazardous situations (such as a state that is not fail-safe, complete loss of control, unsafe shutdown of equipment, loss of propulsion, blackout, etc.).

Corrective action measures should be proposed to correct situations of non-compliance with the single-failure design philosophy.

Undesired Events

The FMEA is to systematically analyse all foreseeable failures and investigate and document if there is a potential for:

- Lost or degraded function beyond acceptable performance criteria of the system under control (i.e., loss of propulsion).
- Hazardous situation (such as a state that is not fail-safe, complete loss of control, unsafe shutdown of equipment).

Note on Fail-safe concept: A fail-safe concept is to be applied to the design of all control systems, manual emergency control systems and safety systems. In consideration of its application, due regard is to be given to the safety of individual machinery, the system of which the machinery forms a part and the vessel as a whole. Below are examples of typical fail-safe states:

System or Component	>	Typical Fail-safe States
- Propulsion speed control	>	Maintain state.
- Controllable pitch propeller	>	Maintain state.
- Propulsion safety shutdown	>	Maintain state and alarm.
- Alarm system	>	Annunciated
- Cooling water valve	>	In most cases, open

Systems or Subsystems

Computer-based systems subject to Classification requirements are to be assigned into the appropriate system category (I, II or III) according to the possible extent of the damage that may be caused by a single failure within the computer-based system.

System Category	*Effects of Failure*
I	Failure will not lead to dangerous situations for human safety, the safety of the vessel and/or threat to the environment (e.g., non essential systems, information, and diagnosis).
II	Failure could eventually lead to dangerous situations for human safety, the safety of the vessel and/or threat to the environment (e.g., Cargo tank gauging system, control systems for auxiliary machinery, main propulsion remote control systems (e.g., the control system from navigation bridge, etc.)).
III	Failure could immediately lead to dangerous situations for human safety, the safety of the vessel and/or threat to the environment.

Category I systems by definition will not cause unsafe operation upon failure, thus there is not a Classification requirement to develop an FMEA. Category II systems may include safety-related functions but there is no specific requirement for the FMEA to be submitted to Classification for review. Note that where independent effective backup or other means of averting danger for the control functions is provided, the system Category III may be decreased to Category II. Also note that in some cases a main propulsion remote control system, automation system or DPS system may be Cat II and because an additional special notation such as ACCU or DPS-2 is selected. In these cases, the FMEA is required to be done as per the requirements of the special notations. Category III systems have the most severe and immediate safety and environmental consequences in case of failure and do require development on an FMEA and submission to Class to support the Classification process. Examples of Category III systems include, but are not limited to:

- Safety system/equipment for main propulsion and electric power generating system associated with propulsion.
- Burner control and safety systems.
- Control system for propulsion machinery or steering gear (e.g., the control system from centralised control station, control system for common rail main diesel engines, etc.); and
- Synchronising units for switchboards.

Failures of components within the computer-based control system are to be addressed in the FMEA, including the interfaces with other systems such as I/O signals. Functional failures of equipment under control shall also be considered as to ascertain the adequacy of the actions of the safety-related functions in case of these failures. In other words, if the control system controls a pumping system, functional failures of the pump and its effects should be analysed to determine what actions the control system takes to mitigate either the likelihood or the effects of the failure. Certain systems with special notations have explicit automation FMEA requirements in the Steel Vessel Rules which are based on this FMEA concept for safety-related function of computer-based control:

- Control system from centralised control station (ACC and ACCU notation).
- Gas turbine safety systems.
- Electronically controlled diesel engines; and
- Gas fuelled engines.

The intent of these requirements is similar to the automation requirement discussed here, but for sake of clarity, they are described in detail in their respective System-Specific FMEA Requirements in Chapter 7.

Modes of Operation

Failure analysis should consider failure scenarios under all potential modes of operations, such as the failure mode, the likelihood and the consequences of the failure scenario vary depending on the operational modes. The modes of operation are specific to the equipment under control. For example, typical modes of operation for a propulsion system will include:

- Start and stop.
- Underway. Consider full range of speeds up to maximum speed and any special issues such as low NOx, fuel optimisation.
- Emergency actions. Consider crash stop from full ahead to full stern, acceleration to maximum rpm (i.e., propellers emerged in heavy seas), etc.

Integrated automation systems can issue commands for the collective benefit and operation of the vessel. This ability to command multiple independent devices could theoretically direct two or more different devices to act in ways that, in combination, could be detrimental. The risk assessment (FMEA or alike) for integrated automation systems is to address possible risks associated with commands given to multiple devices in a way which could lead to a combined undesirable outcome. The risk assessment for integrated automation systems should:

- Consider all possible dangerous outcomes.
- Identify situations that can result in these dangerous outcomes (they could arise from a malfunction of IAS or even from an IAS that works as intended, but their collective operation can create dangerous outcomes); and
- Assess and/or suggest risk controls for situations that could lead to dangerous outcomes, either by software (Cat II/III requirements) or be addressed by means outside of the IAS.

Typical Failures

The lowest level of physical component failure required to be included in the FMEA for control systems is typically at the data acquisition unit level which includes the CPU, its I/O modules, the powering of the device, and any communication between multiple devices. Physical components whose failures should be analysed in the FMEA include:

- CPU microprocessor/PLC.
- Input/output modules.
- Power supplies.
- Electrical power cables.
- Network hubs.
- Buses.
- Communication/data link cables.
- Interfaces/displays.

- Electrical power cables; and
- Others, as applicable.

For functions permitting wireless data communications, consideration is to be given to the possibility of corrupted data and intermittent faults with comparatively long recovery times between interruptions. Software may reside in several different areas of a given system and at multiple levels (i.e., operating system, PLC embedded logic). Software failures are to be considered at a higher level, almost like a system failure; to address how the system can prevent and recover from, for example, malware. With the exception of the ISQM FMEA, a detailed software failure analysis is not usually something considered in FMEA. Classification Rules require the system developer to follow a recognised software quality assurance programme, so it is generally assumed that by the time the software is delivered and installed, it is error-free. Relays, terminal boards, indicator lights, switches, meters, and instruments do not have to be included in the FMEA unless their failure leads to a hazardous situation not discussed with a higher-level component. The risk assessment (FMEA or alike) is also to consider the interfaces in the system, analysing failures related to the i/o such as faults in:

- Signals arriving at IAS (for incorrect data and its effect on system operation); and
- Signals leaving IAS (for combinations that could lead to dangerous outcomes.

Timeline/Team

The sponsor of the FMEA is either the designer of the equipment under control or the system integrator. The optimal time for performing the FMEA is during detailed design when the design is mature enough to have sufficient detail about the system and equipment available, but still enough time is left in the process to include FMEA recommendations in final design and integration process. The FMEA should include participation from subject matter experts in the control systems, operations, design, and relevant vendors.

Verification Programme

The specific goals of the FMEA trial tests include evaluating the:

- Effectiveness of system to identify failures.
- Effect of identified failures on system/equipment.
- Response of safety controls.
- Verification of redundancy as needed.
- Verification of fault tolerance, as needed; and
- Other measures to protect against failure.

Even though there is no general FMEA validation trial for the computer-based systems in this particular requirement, it is to the discretion of the Class Surveyor or Class Plan reviewer to recommend specific testing of FMEA failures for which there is a higher degree of uncertainty. Self-tests such as Hardware-in-the-Loop (HIL) can be substituted, if adequate to prove intent and goals of verification programme. If system failures cannot be replicated (destructive test, safety concerns, etc.), a partial test may be conducted to evaluate functionally and verify existing safeguards, as indicated in the FMEA to detect/prevent/mitigate the failure. Certain systems such as dynamic position systems have computer-based control and are required under special notation requirements to submit an FMEA and FMEA trial plan. These requirements are specified in the DP FMEA requirements (Chapter 7).

Supporting Documents

The FMEA must be submitted to Class for review. In order to conduct a proper review of the FMEA, the Class Plan reviewer needs the following information included with the FMEA report:

- General description of the scope and purpose of the FMEA.
- Dates when the FMEA was conducted.
- FMEA participants.
- Summary of any corrective actions performed or pending as a result of recommendations of the FMEA.
- Detailed description, narrative and drawings of the system under control, and its control system.
- Layout and general arrangement drawings.
- Single line diagrams.
- System design specifications.
- Description of the physical and operational scope and boundaries for the FMEA, including:
  - Functional blocks arranged in a reliability block diagram, showing interactions among the blocks (parallel paths for redundant functional blocks, in series for single blocks).
  - Modes of operation for the system.
- FMEA worksheets, including:
  - Modes of operation.
  - All significant failure modes are associated with each mode of operation.
  - Cause associated with each failure mode.
  - Method for detecting failure has occurred.
  - Effect of the failure on system functionality.
  - Global effect of the failure on other systems, the asset, and HSE.
- Existing controls to prevent and/or mitigate failure; and
- Corrective actions are needed to comply with Classification requirements.

Block diagram showing the system configuration including the user interface, description of hardware specifications, hardware FMEA, fail-safe features, security arrangements, power supply, and independence of systems (control, monitoring, and safety shutdown). For systems where loss of function upon a single failure is not an acceptable option, but redundancy is not possible, further study of non-redundant parts with consideration to their reliability and mechanical protection must be provided. The operations, maintenance, inspection, testing, and emergency manuals are needed for Class to confirm compliance with design and manufacturer's requirements and that the findings of the FMEA have been incorporated into the equipment operating phases. These manuals can all show operational configurations in which the systems redundancy can be bypassed.

Lifecycle Management

Class needs to be notified of any changes made that impact the basis of the Classification requirements. It will be up to the discretion of the Class reviewer whether to require a revised FMEA.

Additional Notes

Nil.

ELECTRONICALLY CONTROLLED DIESEL ENGINES

Purpose of FMEA

The FMEA requirement applies to any electronically controlled diesel engines (e.g., main propulsion engines as well as engines for auxiliary systems such as generators, or other mission-critical functions). The purpose of the FMEA is to demonstrate that a single failure of any component within the electronic control system of the diesel engine will not result in the loss of function for the entire system below Classification performance requirements:

*   For main propulsion, it is not to result in a loss of propulsion performance beyond that allowed per applicable notation requirement; and
*   For auxiliary systems, it is not to result in the loss of the essential service the system provides.

Undesired Events

The FMEA is to evaluate the potential for failure to result in the following consequences:

*   For propulsion system, loss of propulsion performance beyond that allowed per applicable notation requirements; and
*   For auxiliary systems, loss of the essential service provided by system.

Systems or Subsystems

Failures within the following systems are to be addressed in the FMEA:

*   Electronic control systems (for any electronically controlled diesel engine), including but not limited to:
    1. Electronic governor.
    2. Remote telegraph control.
    3. Fuel control (e.g., timing control, electronic over speed, etc.); and
    4. Engine valve control (e.g., exhaust valve, intake valve, etc.).

The FMEA requirement extends to all services which are considered essential to the operation of the engine, therefore, the FMEA of electronically controlled diesel engines shall include the following:

*   For main propulsion engines, include failure analysis of the power management for diesel electric/ multiple engine arrangements; and
*   For auxiliary systems, including failure analysis of the generators (e.g., control for generator speed, isosynchronous, etc.).

Modes of Operation

The FMEA should consider failures of the engine in each of the following modes of operations, in addition to other operations that may be relevant:
- Start and stop.
- Underway. Consider full range of speeds up to maximum speed, and any special issues such as low NOx, fuel optimisation, etc.
- Emergency actions. Consider crash stop, from full ahead to full stern, acceleration to maximum rpm (e.g., propellers emerged in heavy seas), etc.
- Bunkering.
- Manoeuvring in port/coastal passage/ocean passage; and
- Mooring.

Typical Failures

The lowest level of physical component failure required to be included in the FMEA for control systems is typically at the data acquisition unit level which includes the CPU, its I/O modules, the powering of the device, and any communication between multiple devices. Physical components whose failures should be analysed in the FMEA include:

- CPU microprocessor/PLC.
- Input/output modules.
- Power supplies.
- Electrical power cables.
- Network hubs.
- Buses.
- Communication/data link cables.
- Interfaces/displays.
- Electrical power cables; and
- Others, as applicable.

For functions permitting wireless data communications, consideration is to be given to the possibility of corrupted data and intermittent faults with comparatively long recovery times between interruptions. Software may reside in several different areas of a given system and at multiple levels (i.e., operating system, PLC embedded logic). Software failures are to be considered at a higher level, almost like a system failure; to address how the system can prevent and recover from, for example, malware. Classification Rules require the system developer to follow a recognised software quality assurance programme, so it is generally assumed that by the time the software is delivered and installed, it is error-free. Relays, terminal boards, indicator lights, switches, meters, and instruments do not have to be included in the FMEA unless their failure leads to a hazardous situation not discussed with a higher-level component.

Timeline/Team

The sponsor of the FMEA is the equipment vendor. The FMEA should be performed early in the design stage to allow for modifications that may be needed, as well as adequate testing of any of the FMEA assumptions and conclusions. The FMEA should include participation from experts in control systems, design, and operations.

Verification Programme

For electronically controlled diesel engines, integration tests are to verify that the response of the complete mechanical, hydraulic and electronic system is as predicted for all intended operational modes. An FMEA verification programme shall be developed and performed to verify the results of the FMEA. The specific goals of the tests include validating them:

- Effectiveness of system to identify failures.
- Effect of identified failures on system/equipment.
- Response of safety controls; and
- Other measures to protect against failure.

As a minimum, the FMEA verification plan should include evaluating all failures with potential to result in loss of system function. Self-tests such as Hardware-in-the-Loop (HIL) can be substituted, if adequate to prove intent and goals of the verification programme. If system failures cannot be replicated (destructive test, safety concerns, etc.), then existing safeguards must be functionally evaluated and verified, as indicated in FMEA to detect/prevent/mitigate the failure.

Supporting Documents

The FMEA must be submitted to Class for review. In order to conduct a proper review of the FMEA, the Class Plan reviewer needs the following information included with the FMEA report:

- General description of the scope and purpose of the FMEA.
- Dates when the FMEA was conducted.
- FMEA participants.
- Summary of any corrective actions performed or pending as a result of recommendations of the FMEA.
- Detailed description, narrative and drawings of the system and the electronic control system for the diesel engine.
- Layout and general arrangement drawings.
- Single line diagrams.
- System design specifications.
- Description of the physical and operational scope and boundaries for the FMEA, including:
  1. Functional blocks arranged in a reliability block diagram, showing interactions among the blocks (parallel paths for redundant functional blocks, in series for single blocks).
  2. Modes of operation for the system.
- FMEA worksheets, including:
  1. Modes of operation.
  2. All significant failure modes (associated with each mode of operation).
  3. Cause associated with each failure mode.
  4. Method for detecting failure has occurred.
  5. Effect of the failure on system functionality.

6. Global effect of the failure on other systems, the asset, and HSE.
7. Existing controls to prevent and/or mitigate failure.
8. Corrective actions are needed to comply with Classification requirements.

For systems where loss of function upon a single failure is not an acceptable option, but redundancy is not possible, further study of non-redundant parts with consideration to their reliability and mechanical protection must be provided. The operations, maintenance, inspection, testing, and emergency manuals are needed for Class to confirm compliance with design and manufacturer's requirements and that the findings of the FMEA have been incorporated into the equipment operating phases. These manuals can all show operational configurations in which the systems redundancy can be bypassed.

Lifecycle Management

Class needs to be notified of any changes made that impact the basis of the Classification requirements. It will be up to the discretion of the Class reviewer whether to require a revised FMEA.

Additional Notes

Nil.

REMOTE CONTROL PROPULSION [AUTOMATIC CENTRALISED CONTROL (ACC), AUTOMATIC CENTRALISED CONTROL UNMANNED (ACCU), AUTOMATIC BRIDGE CENTRALISED CONTROL UNMANNED (ABCU)]

Rule Requirements

FMEA to be conducted to demonstrate that control, monitoring, and safety systems are so designed that any single failure will not result in the loss of propulsion control, the loss of propulsion or other undesirable consequences. The FMEA report is to be submitted for review.

Purpose of FMEA

The purpose is to demonstrate that any single failure in the control system will not lead to lost or degraded propulsion beyond acceptable performance criteria.

Undesired Events

The FMEA is to analyse all potential failures, one at a time, and investigate and document if there is potential for lost or degraded propulsion beyond acceptable performance criteria.

Systems or Subsystems

* Failures of components within the computer-based control system are to be addressed in the FMEA.
* Failures of the interfaces I/O are to be considered; and
* Functional failures of equipment under control shall also be considered as to ascertain the adequacy of the actions of the safety-related functions upon the controlled equipment failure.

Modes of Operation

The failure analysis should consider failure scenarios under all potential modes of operations, as the failure mode as well as the likelihood and the consequences of the failure scenario vary depending on the operational modes. Typical modes of operation for a propulsion system will include:

* Start and stop.
* Underway. Consider full range of speeds up to maximum speed, and any special issues such as low NOx, fuel optimisation, etc.; and
* Emergency actions. Consider the crash stop, from full ahead to full stern, accelerating to maximum rpm (e.g., propellers emerged in heavy seas), etc.

Typical Failures

The lowest level of physical component failure required to be included in the FMEA for control systems is typically at the data acquisition unit level which includes the CPU, its I/O modules, the powering of the device, and any communication between multiple devices. Physical components whose failures should be analysed in the FMEA include:

- CPU microprocessor/PLC.
- Input/output modules.
- Power supplies.
- Electrical power cables.
- Network hubs.
- Buses.
- Communication/data link cables.
- Interfaces/displays.
- Electrical power cables; and
- Others, as applicable.

For functions permitting wireless data communications, consideration is to be given to the possibility of corrupted data and intermittent faults with comparatively long recovery times between interruptions. Software may reside in several different areas of a given system and at multiple levels (i.e., operating system, PLC embedded logic). Software failures are to be considered at a higher level, almost like a system failure; to address how the system can prevent and recover from, for example, malware. Classification Rules require the system developer to follow a recognised software quality assurance programme, so it is generally assumed that by the time the software is delivered and installed, it is error-free. Relays, terminal boards, indicator lights, switches, meters, and instruments do not have to be included in the FMEA unless their failure leads to a hazardous situation not discussed with a higher-level component.

TIMELINE/TEAM

The sponsor of the FMEA is either the designer of the equipment under control or the system integrator. The optimal time for performing the FMEA is during detailed design when the design is mature enough to have sufficient detail about the system and equipment available, but still enough time is left in the process to include FMEA recommendations in the final design and integration process. The FMEA should include participation from subject matter experts in the control systems, operations, design, and relevant vendors.

Verification Programme

The specific goals of the FMEA trial tests include evaluating the:

- Effectiveness of system to identify failures.
- Effect of identified failures on system/equipment.
- Response of safety controls; and
- Other measures to protect against failure.

Even though there is no general FMEA validation trial for FMEA for ACC or ACCU notation, it is up to the discretion of Class Surveyor or Class Plan reviewer to recommend specific testing of FMEA failures for which there is a higher degree of uncertainty. Self-tests such as Hardware-in-the-Loop (HIL) can be substituted if adequate to prove intent and goals of the verification programme. If system failures cannot be replicated (destructive test, safety concerns, etc.), a partial test may be conducted to evaluate functionally and verify existing safeguards indicated in FMEA to detect/prevent/mitigate the failure.

Supporting Documents

The FMEA must be submitted to Class for review. In order to conduct a proper review of the FMEA, the Class Plan reviewer needs the following information included with the FMEA report:

- General description of the scope and purpose of the FMEA.
- Dates when the FMEA was conducted.
- FMEA participants.
- Summary of any corrective actions performed or pending as a result of recommendations of the FMEA.
- Detailed description, narrative and drawings of the system and its control system.
- Layout and general arrangement drawings.
- Single line diagrams.
- System design specifications.
- Description of the physical and operational scope and boundaries for the FMEA, including:
  1. Functional blocks arranged in a reliability block diagram, showing interactions among the blocks (parallel paths for redundant functional blocks, in series for single blocks).
  2. Modes of operation for the system.
- FMEA worksheets, including:
  1. Modes of operation.
  2. All significant failure modes (associated with each mode of operation).
  3. Cause associated with each failure mode.
  4. Method for detecting failure has occurred.
  5. Effect of the failure on system functionality.
  6. Global effect of failure on health, safety, and the environment.
  7. Existing controls to prevent and/or mitigate failure; and
  8. Corrective actions are needed to comply with Classification requirements.

For systems where loss of function upon a single failure is not an acceptable option, but redundancy is not possible, further study of non-redundant parts with consideration to their reliability and mechanical protection must be provided. The operations, maintenance, inspection, testing, and emergency manuals are needed for Class to confirm compliance with design and manufacturer's requirements and that the findings of the FMEA have been incorporated into the equipment operating phases. These manuals can all show operational configurations in which the systems redundancy can be bypassed.

Lifecycle Management

Class needs to be notified of any changes made that impact the basis of the Classification requirements. It will be up to the discretion of the Class reviewer whether to require a revised FMEA.

Additional Notes

Nil.

GAS TURBINE

Purpose of FMEA

Demonstrate that a single failure in the gas turbine and the gas turbine safety system will not cause a hazardous situation.

Undesired Events

The hazardous situations (or consequence) of concern are:

- Uncontrolled gas ignition; and
- Loss of propulsion.

Systems or Subsystems

- Gas turbines safety system.
- Shutdowns, temperature controls, alarms, starting system, etc.
- Gas turbines (mechanical system).
- Combustion air system.
- Fuel system (gas).
- Compressor.
- Power turbine.
- Lube oil.
- Exhaust gas; and
- Enclosure.

Modes of Operation

- Normal operations.
- Normal startup and shutdown; and
- Emergency shutdown.

Typical Failures

This analysis of the safety systems shall include both:

1. Failures of the components in the safety system (e.g., electronic circuit boards, power supplies, microprocessors/PLCs, memory boards, input/output modules, cables, sensors, etc.).
2. Failures of the mechanical system in order to identify the adequate response of the safety control following the failure.

Typical failures of components and logic within the safety system that need to be analysed in the

FMEA include, but are not limited to:

- Controls and sensors on safety systems related to automatic shutdown, including over speed, high vacuum at compressor inlet, low lubricating oil pressure, low lubricating oil pressure in reduction gear, loss of flame during operation, excessive vibration, excessive axial displacement of rotor, high exhaust gas temperature.
- Controls and sensors related to automatic temperature controls for lube oil/fuel oil/exhaust gas.
- Controls and sensors related to the starting system safety including automatic purging, ignition detection devices, shut off of main fuel valve, and commencing of the purge phase.
- Controls and sensors, including loss of control system power.
- Hand trip gear; and
- Air-intake filters and anti-icing.

Typical failures within the mechanical system (gas turbines) that need to be analysed include:

- Machinery or mechanical components (e.g., failure of rotors, bearing discs, drums, blades, couplings, gears, motor, shaft brake).
- System response to functional failures of supporting utilities.
- Failures at interfaces or interconnections between systems.
- Failures caused by credible external events, especially those that may result in a hazardous situation (e.g., nearby fire, explosion, excessive sea spray, low temperature ambient air, flooding); and
- Others, as applicable.

Particular attention should be paid to the analysis to identify all relevant:

- Common-mode failures and functional dependencies; and
- Hidden failures and their analyses in combination with the initial failure that makes the hidden failure evident (for example, a backup pump does not start upon the failure of the duty pump.

Timeline/Team

- The gas turbine vendor is to coordinate and develop the FMEA and involve relevant operations and subject matter experts; and
- Any corrective actions arising from single failures that lead to the hazardous situation of concern shall be implemented.

Verification Programme

Evaluate the controls and sensors of the safety system by simulating excessive parameter inputs from the sensors and verifying that desired outcome occurs.

Supporting Documents

The FMEA must be submitted to Class for review. In order to conduct a proper review of the FMEA, the Class Plan reviewer needs the following information included in the FMEA report:

- General description of the scope and purpose of the FMEA.
- Dates when the FMEA was conducted.
- FMEA participants.
- Detail narrative of relevant systems, including detail description and narrative of safety system and devices.
- Functional block diagram shows the interactions between systems.
- System design specifications.
- FMEA worksheets, including:
    - All significant failure modes.
    - Cause associated with each failure mode.
    - Common cause failures.
    - Method for detecting failure has occurred.
    - Effect of the failure on propulsion functionality.
    - Effect of the failure on gas fire and explosion.
- Existing controls to prevent and/or mitigate failure; and
- Corrective actions, as needed, to comply with Classification requirements.

For systems where loss of function upon a single failure is not an acceptable option, but redundancy is not possible, further study of non-redundant parts with consideration to their reliability and mechanical protection must be provided. The operations, maintenance, inspection, testing, and emergency manuals are needed for Class to confirm compliance with design and manufacturer's requirements and that the findings of the FMEA have been incorporated into the equipment operating phases.

Lifecycle Management

Any changes in the operating criteria should involve a review of the safety shutdown system FMEA. Class needs to be notified of any changes made that impact the basis of the Classification requirements. It will be up to the discretion of the Class reviewer whether to require a revised FMEA.

Additional Notes

Nil.

REDUNDANT PROPULSION AND STEERING

Purpose of FMEA

Demonstrate that a single failure in any propulsion machine or auxiliary service system will not result in a degradation of propulsion performance.

Undesired Events

The hazardous situation of concern is:

- Loss of propulsion performance.

Systems or Subsystems

- Propulsion machinery.
- Steering machinery.
- Vessel auxiliary services.
- Propulsion control system.
- Electrical generation; and
- Power management system.

Modes of Operation

- Underway.
- Normal startup and shutdown; and
- Emergency shutdown.

Typical Failures

Typical failures that need to be analysed in the FMEA include, but are not limited to:

- Machinery or mechanical components (e.g., failure of gears, motor, brake).
- Propulsion and steering response upon functional failures of supporting utilities.
    1. Fuel oil system.
    2. Lube oil system.
    3. Cooling water.
    4. Compressed air.
    5. Fire protection system.
    6. Ventilation
    7. Emergency shutdowns.
    8. Hydraulic system.
- Control system (e.g., electronic circuit boards, power supplies, microprocessors/PLCs,

memory boards, input/output modules, cables, etc.).
- Failures at interfaces or interconnections between systems.
- Failures caused by credible external events, especially those that may result in eliminating both redundant systems (e.g., fire, explosion, flooding); and
- Others, as applicable.

Particular attention should be paid to the analysis to identify all relevant:

- Common-mode failures and functional dependencies that can nullify both redundant systems; and
- Hidden failures and their analyses in combination with the initial failure that makes the hidden failure evident (for example, a backup pump does not start upon the failure of the duty pump).

Timeline/Team

Two important aspects in this FMEA are common-mode failures and functional dependencies that can defeat both redundant systems. Thus, the optimal time to perform the FMEA is at the system integration phase (shipyard). The propulsion system interfaces with services, controls, etc. will be detailed and the potential common-cause failures will be more evident. If the FMEA for propulsion and steering equipment is performed before the integration with auxiliary systems is finalised, a functional failure analysis of these systems/interfaces should be performed once the design is final in order to evaluate potential common-cause failures.

Verification Programme

There is no specific requirement for evaluating the conclusions of the FMEA. It is at the discretion of the Class Surveyor to test select FMEA failures to validate the system response. Tests of the controls, safety systems, shutdown devices, starting system and emergency power supply, and alarms are to be conducted.

Supporting Documents

The FMEA must be submitted to Class for review. In order to conduct a proper review of the FMEA, the Class Plan reviewer needs the following information included with the FMEA report:

- General description of the scope and purpose of the FMEA.
- Dates when the FMEA was conducted.
- FMEA participants.
- Summary of any risk reducing actions performed or pending as a result of recommendations of the FMEA.
- Detailed description, narrative and drawings of relevant systems:
  1. Redundant propulsion and steering systems.
  2. Propulsion control system.
  3. Power management system.

    4.   Services/utilities relevant to the redundant propulsion and steering
    5.   Vessel automation system.
    6.   Vessel particulars (notations, type of vessel, etc.).
- Layout and general arrangement drawings.
- Single-line diagram.
- System design specifications.
- Functional block diagram showing interactions among the systems, parallel paths for redundant systems, in series for single systems.
- FMEA worksheets, including:
    1.   Modes of operations.
    2.   All significant failure modes are associated with each mode of operation.
    3.   Causes associated with each failure mode.
    4.   Method for detecting failure has occurred.
    5.   Effect of the failure on propulsion functionality.
    6.   Effect of the failure on gas release, fire, and explosion.
    7.   Global effect of failure on other systems, the asset and HSE.
    8.   Existing controls to prevent and/or mitigate failure; and
    9.   Corrective actions are needed to comply with Classification requirements.

For systems where loss of function upon a single failure is not an acceptable option, but redundancy is not possible, further study of non-redundant parts with consideration to their reliability and mechanical protection must be provided. The operations, maintenance, inspection, testing, and emergency manuals are needed for Class to confirm compliance with design and manufacturer's requirements and that the findings of the FMEA have been incorporated into the equipment operating phases. These manuals can all show operational configurations in which the systems redundancy can be bypassed.

Lifecycle Management

Class needs to be notified of any changes made that impact the basis of the Classification requirements. It will be up to the discretion of the Class reviewer whether to require a revised FMEA.

Additional Notes

Nil.

SINGLE POD PROPULSION

Purpose of FMEA

For single pod propulsor design, demonstrate that a single failure will not cause a loss of propulsion and/or steering capability.

Undesired Events

The hazardous situation of concern is the:

- Loss of propulsion and/or loss of steering capability beyond that allowed per applicable notation requirements.

Systems or Subsystems

Machinery and systems associated with podded propulsion:

- Propulsion system (i.e., motor, shafting, brake, propeller).
- Steering system.
- Pod enclosure and shaft seals (to be designed with a double failure criteria).
- Bilge pumping system.
- Pod thrust bearing.
- Lubricating system.
- Cooling and ventilation system.
- Control system for podded propulsor (i.e., hydraulic, pneumatic, electrical, power management); and
- Monitoring and alarm systems, etc. and their subcomponents.

Modes of Operation

- Underway.
- Normal startup and shutdown; and
- Emergency shutdown.

Typical Failures

Typical failures to be considered in the podded propulsion systems:

- Machinery or mechanical components (e.g., failure of gears, motor, brake).
- System response to functional failures of supporting utilities (e.g., lubrication, cooling, hydraulic, pneumatic, electrical, etc.).
- Failures in podded propulsor and steering control system.

- Failures at interfaces or interconnections between systems.
- Failures caused by credible external events, especially those that may result in a hazardous situation (e.g., flooding, fire, and explosion, etc.); and
- Others, as applicable.

Particular attention should be paid to the analysis to identify all relevant:

- Common-mode failures and functional dependencies; and
- Hidden failures and their analyses in combination with the other failure that makes the hidden failure evident (for example, a backup pump does not start upon the failure of the duty pump.

Timeline/Team

- The optimal time for performing the FMEA is at the detailed design when there is enough information about the system and equipment, but still time to include FMEA recommendations in the final design and integration process; and
- The sponsor of the FMEA should be the propulsion system vendors. Needed subject matter experts in design, controls and operations should participate in the FMEA.

Verification Programme

No specific requirement for evaluating the conclusions of this FMEA. Tests of the starting system, controls, safety systems, alarms, shutdown devices and emergency power supply are to be conducted. However, it is at the discretion of the Class Surveyor to test select FMEA failures to validate the system response.

Supporting Documents

The FMEA must be submitted to Class for review. In order to conduct a proper review of the FMEA, the Class Plan reviewer needs the following information included in the FMEA report:

- General description of the scope and purpose of the FMEA.
- Dates when the FMEA was conducted.
- FMEA participants.
- Summary of any corrective actions performed or pending as a result of recommendations of the FMEA.
- Detail narrative of relevant systems.
- System design specifications.
- Layout and general arrangement drawings.
- Description of the physical and operational scope and boundaries for the FMEA, including:
  - Functional blocks arranged in a reliability block diagram, showing interactions among the blocks (parallel paths for redundant functional blocks, in series for single blocks).
  - Modes of operation for the system.
- FMEA worksheets, including:

- Modes of operation.
- All significant failure modes are associated with each mode of operation.
- Cause associated with each failure mode.
- Common cause failures.
- Method for detecting failure has occurred.
- Effect of the failure on propulsion functionality.
- Existing controls to prevent and/or mitigate failure; and
- Corrective actions, as needed for compliance with Classification requirements.

For systems where loss of function upon a single failure is not an acceptable option, but redundancy is not possible, further study of non-redundant parts with consideration to their reliability and mechanical protection must be provided. The operations, maintenance, inspection, testing, and emergency manuals are needed for Class to confirm compliance with design and manufacturer's requirements and that the findings of the FMEA have been incorporated into the equipment operating phases.

Lifecycle Management

Class needs to be notified of any changes made that impact the basis of the Classification requirements. It will be up to the discretion of the Class reviewer whether to require a revised FMEA.

Additional Notes

Nil.

DYNAMIC POSITIONING SYSTEMS (DPS)

Purpose of FMEA

The purpose of the DP FMEA is to verify that system/subsystem and equipment complies with the following failure design philosophy:

- No single failure will lead to a total or partial loss of station keeping capability.
- Effective redundancy exists; and
- Hidden failures are identified and controlled.

Undesired Events

Partial or total station keeping capability. Redundancy is usually needed if a single failure shall not result in a loss of function.

Systems or Subsystems

The systems to be subject to failure analysis include:

1. Main DP systems; and
2. DP auxiliary systems.

Mission-specific systems that interface or have an impact on the DPS and station keeping capability.

Auxiliary Systems

For DPS-2 and DPS-3 notations, failures within the auxiliary systems are to be included in the FMEA. These systems include, but are not limited to, the following:

1. Fuel oil.
2. Lubricating oil.
3. Cooling water.
4. Compressed air.
5. Hydraulic.
6. Pneumatic.
7. Ventilation/HVAC; and
8. Piping system equipment (e.g., purifier, heat exchanger, transfer pump).

System		*Subsystems/Equipment (non-exhaustive list)*	*Components*
Main DP System			
Power System	1	Prime movers with necessary auxiliary systems including piping	Each subsystem has components whose failures should be evaluated. A typical list of representative components is shown below for computer system.
	2	Generators	
	3	Switchboards	
	4	Electrical distribution system (cabling and cable routing)	
	5	Power management if applicable	
Thruster System	1	Thrusters with drive units and necessary auxiliary systems including piping	Each subsystem has components whose failures should be evaluated. A typical list of representative components is shown below for computer system.
	2	Main propellers and rudders if under the control of the DPS	
	3	Thruster control electronics	
	4	Manual thruster controls	
	5	Associated cabling/cable routing	
DP Control System	1	Computer system/joystick system	Computer System: includes programmable electronic devices, associated software, peripherals, interface, etc.
	2	Position reference systems	
	3	DP sensor system	
	4	Display system (operator panels)	
	5	Associated cabling and cable routing	

Mission Specific Equipment

The types of vessels employing DP systems is widely diverse and include, but are not limited to, semi-submersible mobile offshore drilling units (MODU), offshore support vessels and cruise ships. The FMEA is to include failures of mission-specific equipment that interfaces or has an impact on the DP. It is not feasible to show examples of all the potential mission-specific systems, but as an example, typical system-specific operational modes for a pipe-laying vessel may include pipe tensioner and for a MODU, the emergency disconnect and jack-up system may be included, among others. The depth of the analysis with the functional interfaces shall be sufficient to meet the FMEA objectives in the judgment of the FMEA team. For example, to meet the study objectives it may be necessary to analyse components required for product lay such as deployment equipment (tensioner tracks, aligners, etc.), HPUs, electrical supplies and control systems.

DPS-2 and DPS-3

One difference between the DPS-2 and the DPS-3 FMEA lies in how the static components are addressed. Static components in mechanical systems refer to those having parts that normally do not move (e.g., pipes, tanks, vessels, shell-and-tube heat exchanger, manual valves). For electrical/electronic systems, passive components are those that do not require energy to make them work (e.g., electrical cables, resistors, capacitors). For FMEA conducted for DPS-2 vessels, the failure of the static components is not considered unless it is deemed that these components are not adequately protected from damage and their reliability is unproven. For the FMEA conducted for DPS-3 vessels, the failures of the static components must be analysed, as well as two consequences of two other events, fire, and flooding.

Modes of Operation

The FMEA shall consider both global operations at vessel level (e.g., drilling operation for a MODU) and local operations of the DPS (e.g., closed-bus operations). The technical configuration of the DPS system varies for each vessel operation. The FMEA should document the DPS configuration for each vessel operation and analyse failures in all possible modes of operations (global and local systems configurations) in order to identify failures, hazards, and consequences of concern.

Vessel/Mission-Specific Operations

The FMEA is to include failures of mission-specific equipment that interface or have an impact on the DP and station keeping. The DPS FMEA shall analyse failures during all the overall vessel operations as the failure modes and the consequences will vary from one operation to the next. For example, operations to be considered during the DP FMEA of a MODU include:

1. Drilling.
2. Emergency shutdown; and
3. Safe disconnection of risers.

Other operations to consider on DP vessels include, but are not limited to, station-keeping, manoeuvring, weathervaning, product-laying and dredging.

DPS Power Management System Configurations

When there are more configurations for the diesel electric plant design to cope with equipment unavailability (e.g., failures or equipment taken down for maintenance), it is important that all configurations that are possible to be included in DP operations are analysed in the vessel's DP system FMEA to prove that the DP system remains redundant. This should include all open-bus operations and closed-bus operations, etc.

DPS Control System Modes

The DPS control system has several modes of operation, which should be analysed in all possible combinations with the vessel operations and the power configurations. A non-exhaustive list of the DP control modes include:

1. DP control mode (automatic position and heading control).
2. Manual position control mode (centralised manual position control with selectable automatic or manual heading control).
3. Auto track mode (considered as a variant of DP position control, with programmed movement of reference point); and
4. Manual thruster control mode (individual control of thrust (pitch or speed), azimuth, start and stop of each thruster).

Typical Failures

For a DPS-2 or a DPS-3 notation, loss of position is not allowed to occur in the event of a single fault. Single fault includes, but is not limited to, the following:

1. All redundant components, systems, or subsystems.
2. A single inadvertent act of operation (ventilation, fire suppression, ESD) where applicable and if such an act is reasonably probable.
3. Hidden failures (such as protective functions on which redundancy depends) where applicable.
4. Common failure modes.
5. Governor and AVR failure modes where applicable.
6. Main switchboard control power failure modes.
7. Bus-tie protection where applicable.
8. Power management system.
9. DP control system input and output arrangement.
10. Position reference processing.
11. Networks.
12. Communication failure.
13. Automatic interventions caused by external events, when found relevant (e.g., automatic action upon detection of gas); and
14. Fire and flooding events (for DPS-3 notation, and for DPS-2 with additional Fire and Flood Protection notation EHS-F).

For the DPS-2 notation, static components will normally not be considered to fail. For the DPS-2 notation, a single fault includes any active component or system (generators, thrusters, switchboards, DP control computers, sensors, remote-controlled valves, etc.). For the DPS-3 notation, a single fault includes:

1. Items listed above for DPS-2, and any normally static component is assumed to fail.
2. Any components in any one watertight compartment from flooding; and
3. Any components in any one fire subdivision from fire.

DPS Control System Failures

The lowest level of physical component failure required to be included in the FMEA for control systems is typically at the data acquisition unit level which includes the CPU, its I/O modules, the powering of the device, and any communication between multiple devices. Physical components whose failures should be analysed in the FMEA include:

- CPU microprocessor/PLC.
- Input/output modules.
- Power supplies.
- Electrical power cables.
- Network hubs.
- Buses.
- Communication/data link cables.
- Interfaces/displays.
- Electrical power cables; and
- Others, as applicable.

For functions permitting wireless data communications, consideration is to be given to the possibility of corrupted data and intermittent faults with comparatively long recovery times between interruptions. Software may reside in several different areas of a given system and at multiple levels (i.e., operating system, PLC embedded logic). Software failures are to be considered at a higher level, almost like a system failure; to address how the system can prevent and recover from, for example, malware. Classification Rules require the system developer to follow a recognised software quality assurance programme, so it is generally assumed that by the time the software is delivered and installed, it is error-free. Relays, terminal boards, indicator lights, switches, meters, and instruments do not have to be included in the FMEA unless their failure leads to a hazardous situation not discussed with a higher-level component.

DPS-2 and DPS-3 Notation with EHS-series

In addition to the failure modes provided in the Class specific DP Guide, for EHS-series notations, the following failure modes, where applicable, are also to be considered in the FMEA:

1. Operation of protection systems (breakers, bus-ties, etc.) related to short circuits.
2. Severe voltage dips associated with short circuit faults in power plants are configured as a common power system.

3. Failure due to excess and insufficient fuel.
4. Over and under-excitation.
5. Governor and AVR failure modes.
6. Failure modes related to standby start and changeover.
7. Power management failure on load sharing, malfunction, etc.
8. Phase back thrust and large load; and
9. Blackout recovery.

Timeline/Team

- The FMEA should be developed by a group of subject matter experts that includes the following disciplines:
    1. Controls.
    2. Electrical.
    3. Mechanical/marine.
    4. Operations.
- For a new built, the FMEA should be performed early enough to allow design modifications in case there is a finding of non-compliance with Classification failure design principles such as the no-single failure criteria.
- The FMEA is a tool relied on extensively during operations. The DP FMEA must reflect the vessel "as built" condition; therefore, a review and update of the FMEA may be necessary to reflect modifications in the vessel from the time the FMEA was originally developed to the time the vessel is delivered; and
- The FMEA is to be updated after major modifications.

Verification Programme

As per the DP Guide, the FMEA proving trial procedure is to be developed as part of the FMEA study. The objective of the FMEA proving trial is to confirm the FMEA analysis findings and also to confirm that essential functions and features upon which the fault tolerance of the DP system depends are functional as far as it is practical to do so (protections, power management, etc.). The proving trial report is to establish the FMEA test list and the corresponding test procedures including, but not limited to, the following:

1. Purpose of test or failure mode.
2. Vessel and equipment setup.
3. Test method.
4. Expected results.
5. Observed results.
6. Failure detection.
7. Failure effects.
8. Outstanding or resolved action items.
9. Comments; and
10. Witness name, signature, and date for each test.

After completion of DP proving trials, the definitive version of DP FMEA and DP proving trial report, including final analysis/conclusions based on actual results from DP testing, are to be submitted. The FMEA verification plan should include evaluating all failures that have potential to affect functionality (if existing controls fails) as identified in the FMEA. The FMEA that include a criticality ranking are helpful to decide on which items to submit to the proving trial. In general, failures that have potential for major consequences, regardless of the likelihood, should be evaluated. Self-tests such as Hardware-in-the-Loop (HIL) can be substituted if adequate to prove intent and goals of the verification programme. If system failures cannot be replicated (destructive test, safety concerns, etc.), then existing safeguards must be functionally evaluated and verified, as indicated in FMEA to detect/prevent/ mitigate the failure.

Supporting Documents

The FMEA must be submitted to Class for review in support of the Classification process. In most cases the DP Guide will provide a non-exhaustive list of the information that should be provided along with the DP FMEA report. Below is an expanded list of information that, when provided, facilitates a comprehensive review of FMEA by the Class Plan reviewer, thus having an overall positive impact on the Classification review process:

- FMEA report, including FMEA worksheets and FMEA verification plan.
- General description of the scope and purpose of the FMEA.
- Dates when the FMEA was conducted.
- FMEA participants.
- Summary of corrective actions performed or pending as a result of recommendations of the FMEA.
- Detailed description, narrative and drawings of the system functional description and control.
- A description of all the systems associated with the dynamic positioning of the vessel and a functional block diagram showing their interaction with each other. Such systems would include the DP electrical or computer control systems, electrical power distribution system, power generation, fuel systems, lubricating oil systems, cooling systems, backup control systems, etc.
- Layout and general arrangement drawings.
- Brief description of the vessel, vessel's worst-case failure design intent and whether the analysis has confirmed or disproved it.
- Definitions of the terms, symbols, and abbreviations.
- Analysis method and assumptions.
- Single line diagrams.
- System design specifications.
- Cause and effect chart showing the control logic.
- Description of the physical and operational scope and boundaries for the FMEA, including:
    - Functional blocks arranged in a reliability block diagram, showing interactions among the blocks (parallel paths for redundant functional blocks, in series for single blocks).
    - Modes of operation for the vessel, including vessel industrial functions and the DPS configuration for each of the vessel industrial functions.
- FMEA worksheets, including:
    - Modes of operation.

- Description of each physically and functionally independent item and the associated failure modes.
- Cause associated with each failure mode.
- Method for detecting failure has occurred.
- Effects of each failure mode alone on other items within the system and on the overall DPS functionality.
- Existing controls to prevent and/or mitigate failure.
- Conclusions including worst case failure and recommended changes needed to comply with Classification requirements.
- FMEA verification plan; and
- Operations, maintenance, inspection, and testing manuals

The FMEA report is to be updated after major modifications and is to be kept onboard the vessel. For systems where loss of function upon a single failure is not an acceptable option, but redundancy is not possible, further study of non-redundant parts with consideration to their reliability and mechanical protection must be provided. The operations, maintenance, inspection, testing, and emergency manuals are needed for Class to confirm compliance with design and manufacturer's requirements and that the findings of the FMEA have been incorporated into the equipment operating phases. These manuals can all show operational configurations in which the systems redundancy can be bypassed.

Lifecycle Management

The FMEA is to be updated after major modifications and resubmitted to Class to demonstrate that the system is still in compliance with the Class design philosophy. Types of changes that would typically require submission of revised FMEA for Class review include:

- Operating criteria changes.
- System configuration/boundary changes.
- Equipment changes.
- New application/mission.

New FMEA proving trials may be required to validate the conclusions of the updated FMEA.

Additional Notes

Example of helpful columns on a DP FMEA worksheet is shown in the following page.

Table 7.3 Sample DP FMEA Worksheet Template

Operational Mode											
Unit											
Description of Unit	Description of Failure			Effects of Failure			Safeguards		Testing	Corrective Action	Responsible
Function	ID #	Failure Mode	Failure Causes	Detection of Failure	Local	Redundancy Affected?	Global	Prevention of Failure	Mitigation of Effect		

SOFTWARE CONTROL SYSTEM

Purpose of FMECA

For control functions with Integrity Levels (IL) 2 and 3, the FMECA is to demonstrate that a single failure within the control system (software/firmware/hardware) will not cause the failure or degradation of the other functions controlled by the ISQM control system. Corrective actions should be put in place to eliminate any identified potential for losing a control function upon a single failure. System Verification requires a safety analysis report and does not require a FMECA.

System/Global Consequences of Interest

IL 2 and IL 3 correlate to the most severe safety and environmental consequences in case of failure or degradation of any of the functions controlled by the ISQM control system. The consequence of failure of function is defined as follows:

- **IL2 – Critical.** Permanent injury or multiple lost time injuries, mission critical system damage, or significant financial or social loss; and
- **IL3 – Catastrophic.** Loss of human life, loss of asset, loss of system safety or security, or extensive financial or social loss.

System or Subsystems

The FMECA shall analyse failures of the control system (software/firmware/hardware) and as well as failures in the system under control for system functions that have been defined as IL 2 or IL3. A partial listing of recommended ISQM equipment rated IL2 and IL3 follows, based on Section 3, Table 2 of the ISQM Guide. The equipment is selected by the owner based on the owner's risk tolerance:

- Acoustic BOP Control.
- Acoustic DP.
- Ballast Control.
- Ballast Water Treatment.
- Blowout Preventer (BOP).
- Cement Pump.
- Chemical Gas or Oil Processing Separation.
- Draw Works.
- Drilling Control System.
- Drilling Heave Control.
- Drilling Power System.
- Dual Fuel Engine Fuel System.
- Dynamic Positioning.

- Emergency Shutdown (ESD).
- Engine Control System.
- Emergencies Disconnect (EDS).
- Fire and Gas. Fuel Treatment.
- Governor.
- Horizontal Pipe Handling.
- Ice Load Monitoring.
- Lifting Appliances.
- LNG Refrigeration.
- Mud Monitoring Control System.
- Mud Pumps.
- Process Safety Instrumented System.
- Production Subsea ESD.
- Production Subsea Monitoring Thruster.

- Vertical Pipe Handling.
- Vessel Management.
- Vessel Power Management.
- Vessel Stability; and
- Zone Management.

For each system, ISQM is interested in the analysis of:

1. Failures within the control system (software, firmware, and hardware).
2. Failures of the equipment under control (EUC).

Examples of components whose failures must be analysed include, but are not limited to:

- Adapter cards.
- Circuit boards containing solid state devices.
- Circuit drivers.
- Communication boards.
- Computers.
- Electronic circuit boards.
- Ethernet cables.
- Input and output, either remote or local.
- Input/output modules.
- Memory boards.
- Microcontrollers.
- Microprocessors/PLCs.
- Power supplies; and
- Software.

As a rule of thumb, the lowest level of firmware and hardware component failure that should be included in the FMECA is each easily replaceable component. The subsequent section on "Typical Failures for Analysis" addresses the type of software failures to be considered in the analysis. The EUC is the set of all equipment or machinery used for the industrial activities, process, drilling, transportation, etc. (i.e., pumps, compressors, propulsion equipment, power plant, thrusters, cranes, etc.). The EUC can give rise to hazardous events for which the control and safety-related system is required. Therefore, failures within the EUC must be analysed to verify that the control responses and safety-related protection are adequate for all situations. Examples of failures within the machinery/EUC include:

- Functional failures.
- Failures that have the potential to cause unsafe situations.
- Interfaces of EUC and the control system; and
- Failures of sensors in the EUC.

Modes of Operation

The FMECA is to analyse failures occurring under each of the software operating modes and equipment states:

1. Normal.
2. Degraded.
3. Failed (single point).
4. Fail-Safe.
5. Startup; and
6. Shutdown.

In addition, the FMECA is to analyse operational modes specific to the mechanical system under control. Each operational mode may entail different control logic, as well as mode-specific types of failures and consequences.

Typical Failure

Typical software failure modes during each of the operating modes include the following:

1.	Lack of Functionality	Software provides no output or control action not provided when expected.
2.	Improper Functionality	The programmed control system software performs an unexpected action as defined by the operator of the equipment.
3.	Timing	Software event happens too late or too early or control action mistimed.
4.	Sequence	Software event occurs in wrong order or control action with incomplete sequence concept error.
5.	False Alarm/Action	Software detects an error when there is no error or control action provided when not expected.
6.	Faulty Logic and Ranges	Concept error where the software contains incomplete or overlapping logic or control action contains incomplete or overlapping logic.
7.	Incorrect Algorithms	Software computes incorrectly based on some or all inputs or control action is based on wrong computation.
8.	Memory Management	Software runs out of memory or memory leakage or control action stops due to lack of memory.
9.	Interface Failure	Software failed due to failure of hardware interfaces such as power supply, watch dog timer, clocks, reset circuits, etc.
10.	Software Virus	Software did not function on demand due to software virus.

Explicit analysis of the following failure modes for each function should be described in the FMECA:

- Any two opposing functions that canp81 be simultaneously initiated (e.g., valve open and valve close).
- Degraded and/or failed third party components in a Safety Instrumented System (SIS) (as defined by IES61508 and 61511).
- Sensor inputs and outputs are degraded or failed for any function (e.g., components receive a corrupted signal, or a fluctuating signal, a wireless component gets signal interference, etc.).
- Any two degraded or failed complimenting components, including common-cause failures (e.g., the brake off solenoid/air circuit and ACS sensors both fail).
- For redundancy systems with multiple components, consider if one or more redundancy component is degraded or failed.

The failure analysis should include the causes and consequences of the failures, the controls to detect and indicate the failure, and the fail-safe action or logic that will prevent either the cause or the consequence.

Example of a Failure

Failure of Sensors

- Control system function: Monitor the enclosed air pressure in a section.
- Failure: Corrupted signal on pressure transmitter sensor indicating low pressure when in reality there is high pressure
- Consequences: Potential to build up pressure in section without it being noticed. Potential for rupture of section and release of hydrocarbons. Potential for fire.
- Failure Indicating Safeguards: Redundant transmitter with high pressure alarm.
- Safeguards: two out of three voting transmitters, maintenance of the transmitters, pressure relief valve on the section, hydrocarbon detection, area classification, fire protection and firefighting.
- Detectability: How does the failure get detected?

Timeline/Team

- The sponsor of the FMECA is the system provider.
- In addition to the system provider, it is recommended that the owner, operator (driller or crew organisation), ship builder integrator, suppliers involved in integrated software system development, implementation, operation and maintenance, Class ISQM representative and independent auditors are to be present during the FMECA reviews.
- The optimal time for performing the FMECA is at the detailed design when there is enough detail about the system and equipment, but still time to include FMECA recommendations in the final design and integration process. Note that an important focus of the ISQM FMECA is to identify concept errors early to minimise schedule impact.
- Needed subject matter experts in operations, design and relevant vendors should participate in the FMECA.

Verification Programme

No specific requirement for evaluating the conclusions of this FMECA, but it can be utilised for verification plans and testing scenarios. However, it is to the discretion of Class and the owner to evaluate select FMEA failures to validate the system response.

Supporting Documents

The FMECA must be submitted to Class for review. In order to conduct a proper review of the FMECA, the Class reviewer needs the following information included in the FMECA report:

- General description of the scope and purpose of the FMECA.
- Dates FMECA were conducted.
- FMECA participants.
- Summary of any risk-reducing actions performed or pending as a result of recommendations of the FMECA. For large systems, provide a summary of each component. This can also be a table that shows the overall number of failure modes, causes, low-, medium- and high-risk failure modes, etc. and recommendations.
- Detail narrative of relevant systems.
- Design specifications.
- Functional block diagram shows their interaction with each other.
- Unique function identifier for each function for traceability purposes. These are the same identifiers used throughout the project (functional description documents, FMECA, etc.).
- Integrity Level assignments for each function to highlight the importance and criticality of the function to the system. They also represent the severities of the potential consequences. A table is recommended.
- Risk matrix used for critical assessment.
- FMECA worksheets, including:
  - Mode of operation.
  - Description of functions and unique function identifier.
- All significant failure modes.
- Cause associated with each failure mode.
- Common cause failures.
- Method for detecting that the failure has occurred and how the information is passed on to the operator. I should describe the degree of detectability of the function failure. Accompanying tables should be available to explain any value used.
- Effect of the failure upon the system's ability to perform its function – most likely and worst-case outcome.
- Description of major or critical consequences, including potential impacts to personnel and/ operational safety.
- Controls to prevent and/or mitigate failure.
- Risk ranking for each failure mode (severity and consequence).
- Corrective actions – Functions that have been determined to have an unacceptable risk level should describe actions that will be taken to remove or mitigate these risks. For other functions, status on recommendations and action items should be included in the report, if applicable.
- If the system is fault-tolerant (i.e., continue working uninterruptedly after a single hardware failure) but redundancy is not possible, provide further study of non-redundant part with consideration to their reliability and protection mechanisms.

The operations, maintenance, inspection, testing, and emergency manuals are needed for Class to confirm compliance with design and manufacturer's requirements and that the findings of the FMEA have been incorporated into the equipment operating phases.

Lifecycle Management

If the function IL 2 or IL 3 was added after the initial Software Control System FMECA, a new FMECA is to be conducted to address any risks with the new functions and associated Software

Modules. The software FMECA report is a living document, and it should be updated for any subsequent modifications to the system under control as well as to the control system to demonstrate that the system is still in compliance with the ISQM requirements. Types of changes requiring submission of revised FMEA for Class review include:

- Operating criteria changes (this does not include parameterisation changes with the operating limits of the equipment).
- Equipment changes.
- Major software changes for IL2/IL3 system (any functionality updates to IL2 or IL3 functions or software functions that may affect IL2 or IL3 functions are to have updated FMECA.
- Previously unidentified failure modes were discovered; and
- Underlying assumptions are determined to be erroneous (e.g., historical data shows a different likelihood or consequences than the estimated in the analysis).

Additional Notes

Below is a summary of the characteristics expected in a software-focused, top down functional FMECA of IL2 and IL3 designated ISQM control systems for ISQM.

1. *Unique function identifier for each function.* Each system function in the system is to have a unique identifying number for traceability purposes, provided to give clarity and avoid confusion. These are the same identifiers used throughout the project (functional description documents, FMECA, etc.)
2. *Integrity Level assignments for each function.* The IL assignment gives information regarding the importance and criticality of the function to the system. They also represent the severities of the potential consequences. A table is recommended.
3. *Explicit statement for any control system failures that may have an effect upon other IL2 and IL3 functions or components.* This is to provide the designer with a troubleshooting guide to cascading failures.
4. *Corrective actions.* Functions that have been determined to have an unacceptable risk level contain information on actions that will be taken to remove or mitigate these risks. For other functions, statuses on recommendations and action items should be included in the report, if applicable. This provides risk-reducing actions performed or pending as a result of recommendations and negative conclusions in the summary section of the FMECA.
5. *FMECA worksheets for ISQM control software modules assigned IL3 or IL2.* A Control System FMECA is to provide traceability of the Software Modules to the relevant functions. It should provide the failure/function, causes of failures, the effects, method on how and when the failure is detected, the criticality, and what do to reduce the risk of them occurring from a local and global perspective. It should provide the qualitative measure probabilities, the most likely and worst-case outcomes. The criticality should be based on the severity of the outcome and the frequency of the occurrence of the outcome. A risk matrix is the recommended tool for assessment of criticality. Worksheets should include a description of how the failure detection is passed on to the operator. It also may state the degree of detectability of the function. Accompanying documentation should be available to explain any value used. This tells the designer which system, alarm or sensor will detect and react given the failure.
6. *Description of any two opposing functions which can be simultaneously initiated.* This is to

provide the designer with the fail-safe action or logic that will occur when two conflicting inputs impact the functions simultaneously. Example: Valve position switch indicates it is open and closed simultaneously.

7. *Description for the case that the sensor is degraded or failed for any function that controls hardware dependent upon sensors for various positioning or measuring actions.* This is to provide the designer with the consequences of a degraded or defective sensor. Example: If components receive a corrupted signal or fluctuating analogue signal what would be the effect? Wireless components?

8. *Description of the cause and effect of any two degraded or failed complimenting components.* This is to provide the designer with the consequences of any two complimenting components which can fail simultaneously. Example: The brake off solenoid/air circuit and ACS sensor both fail.

9. *Description for the case that the sensor/input is degraded or failed for any function controlling hardware that is dependent upon control valves, motors, wires, pipes, oil/ lubrication, air supply, another component for various positioning or measuring actions.* This is to provide the designer with the causes and consequences of degraded or failed auxiliary components have on various positioning or measuring actions; and

10. *For redundancy systems with multiple components, if one or more redundancy component is degraded or failed, provide a description of the causes and effects.* This is to provide the designer with the causes and effects of systems that have lost their redundancy (fail-safe) components.

Sample FMEA

The appendix shows sample FMEA for Software as per Class ISQM requirements.

Fig. 7.1 Summary of HAZID process for CDS

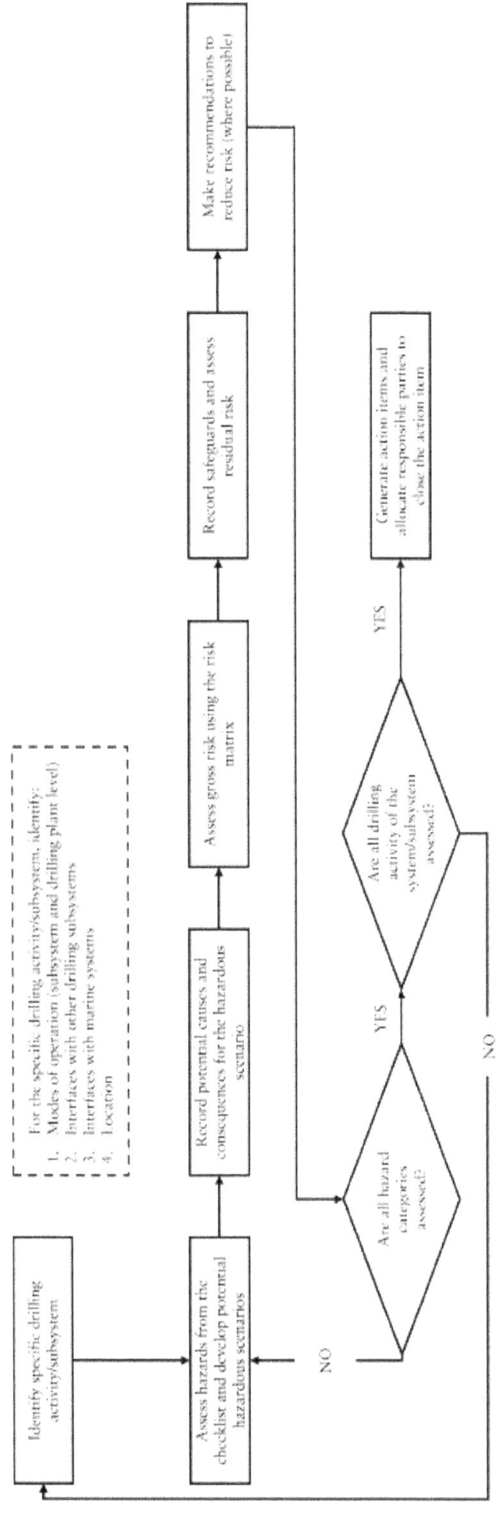

Table 7.4 Sample HAZID Table

HAZID	
Area	Mud process room and Shale Shaker Room
Major Equipment in area	Shale shaker, gumbo conveyor, APV, degasser, flow divider, mud process control room
Modes of Operation	Drilling Operations
Interfaces with other drilling subsystems	Cyber base and control system
Interfaces with marine systems	Electrical system, utility system, sea water, etc.
Reference Docs/Dwgs	GA

Table 7.4 Sample HAZID Table

Hazardous Situations	Causes	Consequences	Safeguards	Recommendations	Comments
Release of H_2S, CO_2, hydrocarbons, etc., into the mud process room and shale shaker room.	Formation gases coming back.	• Potential for injury or loss of life • Fire and Explosion • Equipment failure • Of special concern is the location of the operator's control room within the mud process area.	• H_2S gas detector • Zone certified equipment • Negative pressure in Shaker room • Positive pressure in operator's control room • Air change ration 12x • Crew Operational training H_2S • Fire fighting/detection arrangement • Breathing apparatus or cascade systems as applicable • H_2S operational procedures	Consider a fire and gas and an evacuation study to determine if there is need to relocate operator's control room or add additional safeguards.	The location of the manned operator's control room within the mud process area is of concern as there is a significant likelihood of toxic or flammable gases in area.

JACKING SYSTEMS

Purpose of FMEA

Demonstrate that a single failure of any component will not cause uncontrolled descent of the unit.

Undesired Events

Loss of jacking function > Loss of holding function > Uncontrolled descent.

Systems or Subsystems

The jacking system and holding mechanism comprise the physical scope of the FMEA including:

- Jacking systems.
- Fixation systems.
- Jack case.
- Electrical power distribution system.
- Hydraulic power system.
- Control systems (including programmable systems and their physical components such as programmable logic controllers, network hubs, cards, buses, cabling, encoders, and inter-faces/ displays); and
- Monitoring and alarm systems, etc. and their subcomponents.

Modes of Operation

- Jacking up – raising of the hull, both normal and pre-load.
- Normal holding – i.e., holding of the hull, both normal and pre-load.
- Lowering – lowering of the hull, both normal and pre-load.
- Severe storm holding – elevated and holding under impact load conditions; and
- Emergency operations – jacking up or lowering.

Typical Failures

Typical failures that need to be analysed in the FMEA include, but are not limited to:

- Machinery or mechanical components (e.g., failure of gears, motor, brake).
- Supporting utilities (e.g., loss of hydraulic, electrical systems).
- Control system (e.g., electronic circuit boards, power supplies, microprocessors/PLCs, memory boards, input/output modules, cables, etc.).
- Non-structural static components (e.g., pipes, cables, block valves).
- Critical mechanical components in fixation system.

- Mechanical components of structural members.
- Failures at interfaces or interconnections between systems.
- Failures caused by credible external events (e.g., fire, hurricane); and
- Others, as applicable.

Particular attention should be paid to the Jacking System FMEA:

- The interfaces between the jacking system and other systems that have the potential to affect it.
- The effect of the failure upon the rest of the system's ability to jack the unit including time effects (i.e., if necessary, time is available for manual intervention).
- Yoke and pin type jacking system – pin failure, lifting frame structure failure in the severe storm condition if applicable.
- VFD failure for torque control, speed control and in relationship to brake operation timing, etc.
- Power supply circuit breakers/relays, wiring and control interlocks.
- Punch-through and control panel location on open deck.
- Anti-collision between the hull and the spud-can.
- Critical monitoring failures such as RPD, speed, overload, overcurrent, over-torque, and safety system actions based on the normal operation of these monitoring; and
- Rack chock/fixation system failure.

Timeline/Team

- The optimal time for performing the FMEA is at the detailed design stage when there is enough detail about the system and equipment, but still time to include FMEA recommendations in the final design and integration process; and
- Needed subject matter experts in operations, design and relevant vendors should participate in the FMEA.

Verification Programme

There are no specific Classification requirements for testing to validate FMEA conclusions for jacking systems. However, it is at the discretion of the Class Surveyor or Class Plan reviewer to recommend specific testing of FMEA failures for which there is a higher degree of uncertainty.

Supporting Documents

The FMEA must be submitted to Class for review. In order to conduct a proper review of the FMEA, the Class Plan reviewer needs the following information included in the FMEA report:

- General description of the scope and purpose of the FMEA.
- Dates when the FMEA was conducted.
- FMEA participants.
- Summary of any risk-reducing actions performed or pending as a result of recommendations

of the FMEA.
- Detail narrative of relevant systems.
- System drawings, layout and general arrangement, Process Flow Diagrams, etc.
- Functional block diagram showing systems interactions, parallel paths for redundant systems, in series for single systems.
- Description of the system boundaries, both physical and operational.
- System design specifications.
- FMEA worksheets, including:
  1. Modes of operations.
  2. All significant failure modes associated with each operational mode.
  3. Cause associated with each failure mode.
  4. Common cause failures.
  5. Method for detecting failure has occurred.
  6. Effect of the failure on the system, including its ability to conduct its function.
  7. Global effect of failure on other systems, the asset and HSE; and
  8. Controls to prevent and/or mitigate failure.

Where redundancy is needed to prevent "uncontrolled descent" but is not possible, further study of the non-redundant parts must be conducted to provide evidence of their reliability and mechanical protection. The operations, maintenance, inspection, testing, and emergency manuals are needed for Class to confirm compliance with design and manufacturer's requirements and that the FMEA findings have been incorporated into equipment operating phases.

Lifecycle Management

If any subsequent modifications to the jacking/holding system are conducted, risk studies are to be updated to demonstrate that the system is still in compliance with the Class no-single failure criteria. Types of changes requiring submission of revised FMEA for Class review include:

- Operating criteria changes.
- System configuration/boundary changes.
- Equipment changes.
- New application.
- Revalidation of previous FMEA from the vendor.

Additional Notes

Nil.

SUBSEA HEAVY LIFTING

Purpose of FMEA

Demonstrate that a single failure during sub-sea heavy lifting will not cause loss of control of load, structural damage, or other hazardous situations such as fire and explosion, pollution, or injury.

Undesired Events

Loss of control of load, structural damage, or other hazardous situations such as fire and explosion, pollution, or injury.

Systems or Subsystems

The physical scope of the heavy lifting system for the subsea includes cranes, winches, cylinders, accumulators, active/passive heave compensation, electric circuits and control system.

Modes of Operation

The FMEA is to consider failure during all foreseeable subsea lifting modes of operations, including the use of active heaving compensation, passive heaving compensation, constant tension, or any combination of these systems.

Typical Failures

Typical failures that need to be analysed in the FMEA include, but are not limited to:

- Electric failures.
- Hydraulic failures.
- Active/passive compensation failure.
- Mechanical failures.
- Failure of crane power supply from vessel.
- Failures of lifting active equipment components.
- Failures of passive components such as pipes, cables, block valves, etc.
- Failure of mechanical components of structural members.
- Failures of control systems at I/O level (e.g., sensors, power supply, PLC, cables, etc.).
- Failures at interfaces or interconnections between systems.
- Failures caused by credible external events (e.g., fire, hurricane); and
- Others, as applicable.

Particular attention should be paid to the FMEA to the interfaces between the heavy lift system and other systems that have the potential to affect it.

Timeline/Team

- The optimal time for performing the FMEA early design is at the vendor stage when there is still time to include FMEA recommendations in the final design and integration process.
- Ideally, subject matter experts in operations, design and relevant vendors should participate in the FMEA; and
- The FMEA report shall be submitted to Class with approval drawings.

Verification Programme

There are no specific Classification requirements for testing to validate the FMEA conclusion. However, it is at the discretion of the Class Surveyor or Class Plan reviewer to recommend specific testing of FMEA failures for which there is a higher degree of uncertainty.

Supporting Documents

The FMEA must be submitted to Class for review. In order to conduct a proper review of the FMEA, the Class Plan reviewer needs the following information included in the FMEA report:

- General description of the scope and purpose of the FMEA.
- Dates when the FMEA was conducted.
- FMEA participants.
- Summary of any risk-reducing actions performed or pending as a result of recommendations of the FMEA.
- Description of active or passive heavier system functions, control, and reliability.
- Description of vessel power supply system.
- System drawings, including layout/general arrangement, schematics.
- Functional block diagram showing systems interactions, parallel paths for redundant systems, in series for single systems.
- Description of the system boundaries, both physical and operational.
- System design specifications.
- FMEA worksheets, including:
  1. Modes of operations.
  2. All significant failure modes associated with each operational mode.
  3. Cause associated with each failure mode.
  4. Common cause failures.
  5. Method for detecting failure has occurred.
  6. Effect of the failure upon the system.
  7. Global effect of failure on other systems, the asset and HSE; and
  8. Controls to prevent and/or mitigate failure.

The operations, maintenance, inspection, testing, and emergency manuals are needed for Class to confirm compliance with the design and the manufacturer's requirements and that the FMEA findings have been incorporated into the equipment operating phases.

Lifecycle Management

If any subsequent modifications to the vessel or the cranes are conducted, risk studies are to be updated to demonstrate that the system is still in compliance with the Class design philosophy. Types of changes requiring submission of the revised FMEA for Class review include:

- Operating criteria changes.
- System configuration/boundary changes.
- Equipment changes.
- New application; and
- Revalidation of previous FMEA from vendor.

Additional Notes

Nil.

DRILLING SYSTEM/SUBSYSTEMS/INDIVIDUAL EQUIPMENT

Purpose of FMEA

To verify that the system/subsystem and equipment comply with failure design philosophy: no single failure will lead to a hazardous situation to people, environment, or equipment; and there are at least two protections in place to prevent the hazardous event.

Undesired Events

1. Injury, pollution, or equipment damage; and
2. Loss of functionality is also a consequence of concern for well control systems and drill string systems with active heavily compensation (AHC).

Safe shutdown is usually sufficient to prevent hazardous situations, and redundancy is usually needed if a single failure shall not result in a loss of function.

Systems or Subsystems

A typical list of systems contained in a drilling unit is included in the CDS Guide and is reproduced below. All major well control, drilling support and universal support systems need an FMEA in order to validate the failure of the design philosophy. Simple systems can be validated without an FMEA as indicated in the section below:

Well Control	*Drilling Support*	*Support*
Blowout Preventers	Derrick/Mast	Storage
Lower Marine Riser Package	Mud Conditioning/Return	Nitrogen Generation/Charging
Marine Drilling Risers	Cementing	
Diverter Systems	Conductor Tension System Unit	Hydraulic and Pneumatic Systems
Choke and Kill Systems	Drill String Compensators	Utilities and Instrument Air
Well Circulation	Riser Tensioning	Chemical Injection Unit
Auxiliary Well Control Equipment	Hoisting Equipment	Sea Water
Secondary Well Control Systems	Lifting/Cranes	
Emergency Well Control System	Pipe/Riser/BOP Handling	
	Rotary Equipment	
	Hydraulic Power Units	
	Well Test	
	Hydrocarbon Disposal	

Note: For hydrocarbon systems, HAZOP is a viable alternative to FMEA.

Systems Exempt from FMEA

Complex drilling systems need an FMEA in order to rigorously evaluate compliance with Classification failure design principles, but for simple systems, standard engineering methods suffice to demonstrate compliance with design principles. Simple systems are those for which the following statements hold true:

- Manual control only – could be hydraulic, pneumatic, electrical, etc. (e.g., on/off switches, manual lever).
- Do not have any PLC or computerised control system.
- Operator cannot alter control function.
- Go mechanically to a safe state in all failure modes.
- Failure of control will not impact another machine; and
- The system will not be impacted by other machine failure.

Examples of simple systems are air-powered winch (hoisting equipment) and joystick-controlled equipment.

Modes of Operation

A system normally has multiple modes of operation, and each mode can present distinct failure scenarios. The FMEA shall consider both global operations at installation or vessel level (i.e., drilling operation), and local operations of the system/equipment which is part of the scope of the FMEA as relevant. The FMEA must analyse all the modes of operations (global and local) to identify failures, hazards, and consequences of concern. The following list gives an example of modes of operations that should be explicitly considered on the FMEA, both for failure modes that are operation-specific, as well as resultant consequences that vary depending on the operations. Global MODU operations in a MODU may include the following:

- Transit between wells (transit/operations/preparations in between drilling two wells).
- Pre-drilling.
- Deployment.
- Upon latching.
- Normal drilling and well control.
- Simultaneous operations.
- Emergency operations.
- Transient operations; and
- Subsystem-specific operational modes.

It is not feasible to show examples of all the local operational modes for all systems, but as an example, typical system-specific operational modes for the Active Heave Compensated draw works may include tripping; weighting on cement (WOC); active heavy compensation (during drilling); lock-to-bottom (while running BOP); and slip and cut.

Typical Failures

Mechanical Designs (Functional FMEA)
Typical failures that need to be analysed during failure analysis of a system/subsystem or equipment include, but are not limited to:

- Machinery or mechanical components (e.g., pump, compressor).
- Components within the control system.
- Non-structural static components (e.g., pipes, cables, block valves).
- Mechanical components of structural members.
- Interconnections between systems; and
- Credible external events (e.g., fire, hurricane).

Programmable Computer Controls (Component-Level FMEA)

The systems with programmable computer controls should include a failure analysis of the components of the control system. The lowest level of physical component failure required to be included in the FMEA for control systems is typically at the data acquisition unit level which includes the CPU, its I/O modules, the powering of the device, and any communication between multiple devices. Physical components whose failures should be analysed in the FMEA include:

- CPU microprocessor/PLC.
- Input/output modules.
- Power supplies.
- Electrical power cables.
- Network hubs.
- Buses.
- Communication/data link cables.
- Interfaces/displays.
- Electrical power cables; and
- Others, as applicable.

For functions permitting wireless data communications, consideration is to be given to the possibility of corrupted data and intermittent faults with comparatively long recovery times between interruptions. Software may reside in several different areas of a given system and at multiple levels (i.e., operating system, PLC embedded logic). Software failures are to be considered at a higher level, almost like a system failure; to address how the system can prevent and recover from, for example, malware. Classification Rules require the system developer to follow a recognised software quality assurance programme, so it is generally assumed that by the time the software is delivered and installed, it is error-free. Relays, terminal boards, indicator lights, switches, meters, and instruments do not have to be included in the FMEA unless their failure leads to a hazardous situation not discussed with a higher-level component.

Timeline/Team

- Individual systems and equipment FMEA are ideally performed by the vendor.
- Timing early enough to allow design modifications in case there is a finding of non-compliance with Classification failure design principles such as the no-single failure criteria.

Verification Programme

As per the CDS Guide, a Validation Programme is to be developed and performed to verify selected/critical results of the FMEA/FMECA/Risk Assessment of the Equipment. The specific goals of the tests are to validate the:

- Effectiveness of system to identify failures.
- Effect of identified failures on system/equipment.
- Response of safety controls; and
- Other measures to protect against failure.

This testing programme is subject to Class Engineering Approval and Survey Witness. The Equipment Possessing Controls Systems subject to Class Engineering Approval and Unit Certification are required to undergo Validation Testing with Survey witness and acceptance.

1. FMEA/FMECA validation is to be conducted at the vendor's plant, in accordance with the CDS Guide.
2. When final testing requires assembly and installation onboard facility, it may not be possible to perform all required testing at vendor's plant. In this case, FMEA/FMECA validation is being conducted as part of the system integration testing (SIT) during commissioning, in accordance with the CDS Guide; and
3. The test programme results are to be documented to confirm that the FMEA/FMECA conclusions are valid. When equipment control systems possess mitigating design barriers which prevent single-point-failures (of controls systems) from cascading to unsafe events, the appropriate function response of an applicable barrier to perform as intended is the critical result subject to validation testing. Critical Results subject for validation testing should NOT be directly correlated to a client's chosen Criticality ranking of an FMECA. While a FMECA Criticality ranking can be aligned toward this criterion to help identify such mitigating barriers, a direct correlation of clients chosen criticality ranking method for identifying CDS Critical results should NOT be done until the designers' chosen methodology is verified to identify alignment.

Critical Results are defined as those areas of a mechanical or control system which has mitigating barriers(s) to prevent the occurrence of a hazardous situation. The verification programme is to validate that upon the specified failure, a minimum of one mitigating barrier performs as intended in order to prevent the occurrence of a hazardous situation. Examples of such mitigating barriers are hydraulic load holding valves, alarms, sensors, etc. Selected Results are defined as those results for which there is reasonable uncertainty or disagreement of the FMEA assumptions. During the FMEA, uncertain assessments results are to be identified and discussed with the designer at time of Design Review for resolution. If a resolution is not achieved, these items should be included in the verification programme. Examples of areas where inadequate data may be available to perform definitive analysis, thus should be part of the verification programme include the behaviour of interlocks that may inhibit operation of essential systems.

Technical Notes

Successful FMEA/Risk Validation Testing will verify the appropriateness of design, but one test cannot be extended for Unit Certification for additional equipment assemblies. The reason for this is that construction, assembling, and component integrity can vary with each manufacturing which in turn may result with a needed barrier not performing as intended. Thus, as required by Chapter 4, Table 4.1 and Chapter 7. Validation Testing is to be conducted for each manufactured unit for all equipment/control systems requiring Engineering Design approval and Unit Certification as applicable per the CDS Guide. If a Barrier is identified to be tested for Validation that cannot be tested without damage to the unit, the requirement for repeating the testing on each subsequent manufactured unit may be waived on a case-by-case basis at the discretion of the reviewing Class Engineering Office, subject to a successful outcome of the testing of the first unit. Self-tests such as Hardware-in-the-Loop (HIL) can be substituted if adequate to prove intent and goals of verification programme. When a full test of a system failure cannot be performed (destructive test, safety concerns, etc.), its functional testing must be conducted as far as mutually agreed. In addition, the barriers to detect/prevent/mitigate failure, as indicated in the FMEA, must be functionally evaluated.

Supporting Documents

The FMEA must be submitted to Class for review. In order to conduct a proper review of the FMEA, the Class Plan reviewer needs the information listed in the CDS Guide. This essential documentation includes the following:

- FMEA report, including FMEA worksheets and FMEA verification plan.
- General description of the scope and purpose of the FMEA.
- Dates when the FMEA was conducted.
- FMEA participants.
- Summary of any risk-reducing actions performed or pending as a result of recommendations of the FMEA.
- Detailed description of the system and its function(s).
- Drawings of the system and control.
- Layout and general arrangement drawings.
- Single line diagrams.
- System design specifications.
- Cause and effect chart showing the control logic.
- Description of the physical and operational scope and boundaries for the FMEA, including:
- Functional blocks arranged in a reliability block diagram, showing interactions among the blocks (parallel paths for redundant functional blocks, in series for single blocks).
- Modes of operation for the system.
- FMEA worksheets, including:
- Modes of operation.
- All significant failure modes (associated with each mode of operation).

- Cause associated with each failure mode.
- Method for detecting the failure has occurred.
- Effect of the failure on system functionality.
- Global effect of failure on health, safety, and the environment.
- Existing controls to prevent and/or mitigate failure.
- Corrective action required to comply with Classification requirements.
- FMEA Verification plan; and
- Operations, maintenance, inspection, and testing manuals.

For systems where loss of function upon a single failure is not acceptable, but redundancy is not possible, provide a further study of non-redundant parts with consideration to their reliability and mechanical protection. The operations, maintenance, inspection, testing, and emergency manuals are needed for Class to confirm compliance with design and manufacturer's requirements and that the findings of the FMEA have been incorporated into the equipment operating phases. These manuals may also highlight operational configurations in which the systems redundancy could be bypassed.

Lifecycle Management

Risk studies* results are to be maintained by the Owner of the drilling unit. If any subsequent modifications to the drilling system, subsystems, equipment, or components are conducted, risk studies are to be updated to demonstrate that:

1. Any hazards derived from the modifications have been mitigated; and
2. The system is still in compliance with the Class no-single failure criteria.

These risk studies shall be resubmitted to Class after the modification.

* Risk Studies refers to the analysis of the integrated systems (e.g., HAZID) and systems/equipment analysis (e.g., FMEA)

Additional Notes

- Equivalent risk methods can be used as long as they fulfil the intent of the FMEA.
- FMEA recommended it because it is well suited for mechanical systems; and
- The CDS Guide uses the term FMEA/FMECA. The FMECA is an FMEA that includes a critical assessment of the failure and its consequences. This criticality assessment is not needed for Class review purposes of CDS equipment. For the purposes of the CDS Guide, an FMEA will always be sufficient. The FMEA is to prove compliance with key principles of the Class design philosophy. These principles need to be complied with, regardless of the assessed criticality.

INTEGRATED DRILLING PLANT

Purpose of HAZID

The purpose of the Hazard Identification (HAZID) study is to identify the major hazards to people, environment and equipment associated with the integrated drilling systems/drilling plant and verify the adequacy of the risk control measures. The HAZID study shall focus on identifying hazardous situations* originating from:

1. Arrangement, location and general layout of drilling systems, subsystem, equipment.
2. Integration and interconnections between the different drilling systems, subsystems, and equipment.
3. Interfaces with drilling support systems, utilities, and marine systems; and
4. Various operating modes.

Undesired Events

Harm to:

1. Human safety.
2. The environment.
3. Equipment.

Systems or Subsystems

In the CDS Guide, integrated drilling systems refers to the collective of systems, subsystems and equipment related to well control, drilling support and support systems installed on the drilling unit. The integrated drilling systems to be analysed for major hazards include all the systems identified in the CDS Guide, such as:

- Well control.
- Marine drilling riser.
- Drill string compensation.
- Riser tensioning.
- Drilling fluid system – mud conditioning and circulation.
- Bulk storage and transfer.
- Drilling.
- Hoisting, lifting, tubular handling.
- Mechanical load-bearing equipment.
- Well testing, completion, and servicing system.
- Control and monitoring system for above.
- Drilling support; and
- Others, as appropriate.

The detailed analysis of these systems will be accomplished via the FMEA of CDS systems/ subsystems and components. The layout and physical location of systems is important to consider in the HAZID as it can be the cause of hazardous situations. The areas below should be looked into during the HAZID:

- Derrick and drill floor and surrounding area.
- Moon pool area.
- Driller's control cabin.
- Central controls area.
- Local electrical room.
- Choke and kill manifold area.
- BOP and Christmas tree storage area.
- BOP central control unit (CCU).
- Hydraulic Power Unit (HPU) area (BOP/ Diverter, Drilling system).
- Subsea engineer's room.
- Well test area.
- Drain, drain collection.
- Mud pit area.
- Mud pump room.
- Cementing unit area.
- Mud treatment.
- Mud lab.
- Cuttings/solid disposal and storage.
- Bulk material transfer area and storage.
- Storage areas (e.g., pneumatic bottles, tools, equipment).
- Drilling switchboard room.
- Tubular handling areas.
- Tubular storage (forward and aft of moon pool).
- Overhead cranes.
- Pipe, cable, utility trunk.
- Mud gas separator, choke, and kill manifold area.
- Flare boom.
- Remote-operated vehicle (ROV) storage and launch area.
- Utilities and drilling support system areas (e.g., nitrogen generation, chemical injection); and
- Others, as appropriate.

Interfaces with Marine Systems

Marine systems are not covered in the CDS Guide but their interfaces with the drilling plant must be addressed in the HAZID study, as they can result in hazardous situations to people, environment, or equipment. The integrated system HAZID study must analyse the potential issues/ hazards from interfaces with marine areas such as:

- DP system/mooring.
- Supply vessels.
- Laydown areas.
- Air compressor room.
- Electrical supply (e.g., high voltage and drilling switchboard rooms).
- Drain collection and cleaning.
- Exhaust systems.
- Heat tracing.
- Firefighting systems.
- HVAC.
- Other utilities (e.g., potable water, control air, cooling media); and
- Other emergency systems.

Modes of Operation

A system normally has multiple modes of operation, and each mode can present distinct failure scenarios (failure modes, hazards, consequences). HAZID must analyse all the modes of operations to identify hazards and consequences of concern. The HAZID shall consider both global operations at installation or vessel level (i.e., drilling operation), and local operations of the systems in question. The following list gives examples of modes operations that should be explicitly considered on the HAZID study for the hazards that are operation specific. Other operations may need to be included, as applicable. Typical **global** and **local** operations of a drilling unit that should be considered in the HAZID are listed below:

Operation	*Typical Activities Occurring During Global Operations of Drilling Unit*
Before drilling	Preparations of tubulars, drilling fluids, etc. prior to drilling such as: • Complete testing per API 53 requirements for all well control equipment. • Functional test other surface equipment as necessary. • Pre-charge BOP accumulator. • Conduct well control drill and emergency drill. • Make a drill stand.
Transit between wells	Operations and preparations in between drilling two wells such as: • Inspect retrieved BOP/LMRP equipment, riser, C&K line diverter equipment etc. • Perform necessary maintenance and inspection on all well-controlled equipment.
Deployment and retrieval	• Deployment of rig, BOP, LMRP, etc. • Drill bare hole using drill string. • Install outer conductor casing, cementing. • Lower BOP stack assembly using riser. • Deploy C&K hoses, flow line, hot line, mud boost line etc. • Line up diverter for shallow drilling. • Retrieval of BOP, LMRP, riser, and moving to designated stored position.
Upon latching	Latching of BOP, LMRP, riser, etc. Once latched perform all testing per API 53 requirement, function evaluate all functions for BOP, C&K, diverter, etc.
Normal drilling and well control	• Drilling the surface hole and subsequent section of well per drilling plan, tripping out, casing, cementing, testing, completing, etc. • Well-kicks, unexpected pressures, gas cut mud, etc.
Simultaneous operations	• Dual-derrick operations – such as drilling on derrick, making casing string on other. • Two or more operations being conducted concurrently such as production, construction, heavy lifts, barges, rig movement, emergency response, etc.

Operation	*Typical Activities Occurring During Global Operations of Drilling Unit*
Emergency Operations	Weather-related emergencies, loss of well control, collision, EDS, loss of DP, loss of power, riser string component breakage, tensioning system failure, marine event such as loss of ballast control, AHD draw work failure etc.
Special and transient operations	Startup, shutdown, riding storms, wireline, jarring, well test, well completion.
Subsystem-specific operations	It is not feasible to show examples of all the operational modes for all systems, but as an example, typical system-specific operational modes for the Active Heave Compensated (ACH) draw works may include, but are not limited to the following: • Tripping-in and tripping-out. • Active heavily compensation (during drilling). • Lock-to-bottom (while running BOP or well testing). • Riding a storm attached. • Others as applicable.

Typical Failures

HAZID workshop shall identify potential hazardous situations and verify that adequate risk control measures are provided to address:

- Loss of control of load/dropped objects.
- Release of hydrocarbons.
- Release of H2S.
- Release of toxic chemicals.
- Fire and explosion.
- Pollution of the sea.
- Loss of well control/blowouts.
- Release of pressurised fluids.
- Loss of system function.
- Loss of station-keeping.
- Loss of stability/buoyancy.
- Structural damage.
- Damage of subsea equipment.
- Equipment collisions.
- Injury due to interface with machinery.
- Blackout.
- Environmental events.
- Failure of mechanical components under stress.
- Domino effects/impairment of emergency response equipment/activities (Not to be considered on its own, but in conjunction with each hazardous event).

The integrated drilling plant HAZID study shall look into the following items:

- Safety of personnel and operation.
- Separation of non-hazardous areas from classified hazardous areas.
- Separation of fuel and ignition source as far as practical.
- Minimising the spread of flammable liquids and gases
- Minimising consequences of fire and explosions.

- Minimising probability of ignition.
- Providing for adequate arrangements for escape and evacuation.
- Facilitating effective emergency response.
- Minimising the likelihood of uncontrollable releases of hydrocarbon.
- Minimising dropped object hazards to personnel, equipment (on installation and subsea) and structure.
- Protection of critical systems, subsystems, equipment, and/or components from damage during drilling operation, such as:
  - Electrical cables and cableways.
  - Well-controlled equipment.
  - Exhaust ducting and air intake ducting.
  - Control and shutdown systems.
- Preventing fire escalation and equipment damage.
- Fire/gas detection and firefighting equipment arrangement so that they are protected from damage during drilling operations; and
- Equipment provides access for inspection and servicing and safe means of egress from all machinery spaces.

Timeline/Team

The HAZID for the integrated drilling plant should be performed early enough in the design stage to facilitate implementation of recommendations that may involve a design change. The HAZID sponsors are typically the:

- Overall Drilling Plant integrator (shipyard); and
- Vessel owner.

Subject matter experts in appropriate disciplines are key to the successful identification of hazards and appropriate development of the HAZID scenarios. Additional participants in the HAZID include subject matter experts from:

- Owner.
- Operations (e.g., drilling, pipe handling, mud, process).
- Integrator or party responsible for integration (e.g., shipyard or major equipment supplier).
- Shipyard.
- Major equipment vendors.
- Control system integrator.
- Classification subject matter experts.
- Subsea.
- Marine.
- Mechanical/piping.
- Electrical/Instrumentation and control.
- Safety/HSE.
- HAZID team facilitator and HAZID scribe with extensive experience facilitating/recording HAZID studies.

Verification Programme

Not applicable to Integrated Drilling Plant HAZID studies.

Supporting Documents

The HAZID report must be submitted to Class for review. In order to conduct a proper review, the Class Plan reviewer needs the following information:

- Completed HAZID report including the following:
  - Description of objective and scope.
  - Description of hazard identification technique used.
  - List of team members who participated in the workshop.
  - Description of each action item and responsible party to close the item, as identified by the team during the workshop.
- Completed HAZID workshop record describing the potential accident scenarios discussed by the team as well as corrective actions.
- Detailed description, narrative and drawings of the system functional description and control.
- Layout and general arrangement drawings.
- System design specifications.
- Cause and effect chart showing the control logic.
- Description of the physical and operational scope and boundaries for the risk study.
- Operations, maintenance, inspection, and testing manuals.

The operations, maintenance, inspection, testing, and emergency manuals are needed for Class to confirm compliance with the design and the manufacturer's requirements and that the findings of the risk study have been incorporated into the equipment operating phases.

Lifecycle Management

Risk studies* results are to be maintained by the Owner of the drilling unit. If any subsequent modifications to the drilling system, subsystems, equipment, or components are conducted, risk studies are to be updated to demonstrate that any hazards derived from the modifications have been mitigated. These risk studies shall be resubmitted to Class after a modification.

** Risk Studies refers to Integrated systems (e.g., HAZID) and Systems/Equipment (e.g., FMEA).*

Additional Notes

- The requirements for FMEA for the individual CDS systems are discussed in Chapter 7; and
- Note that a drilling plant HAZID is always a requirement of the CDS Guide, even if the drilling plant is a standard design. For previously approved designs for which a risk assessment exists, the HAZID should be updated. If no risk assessment exists, complete HAZID must be conducted.

Fig. 7.1 Summary of HAZID process for CDS

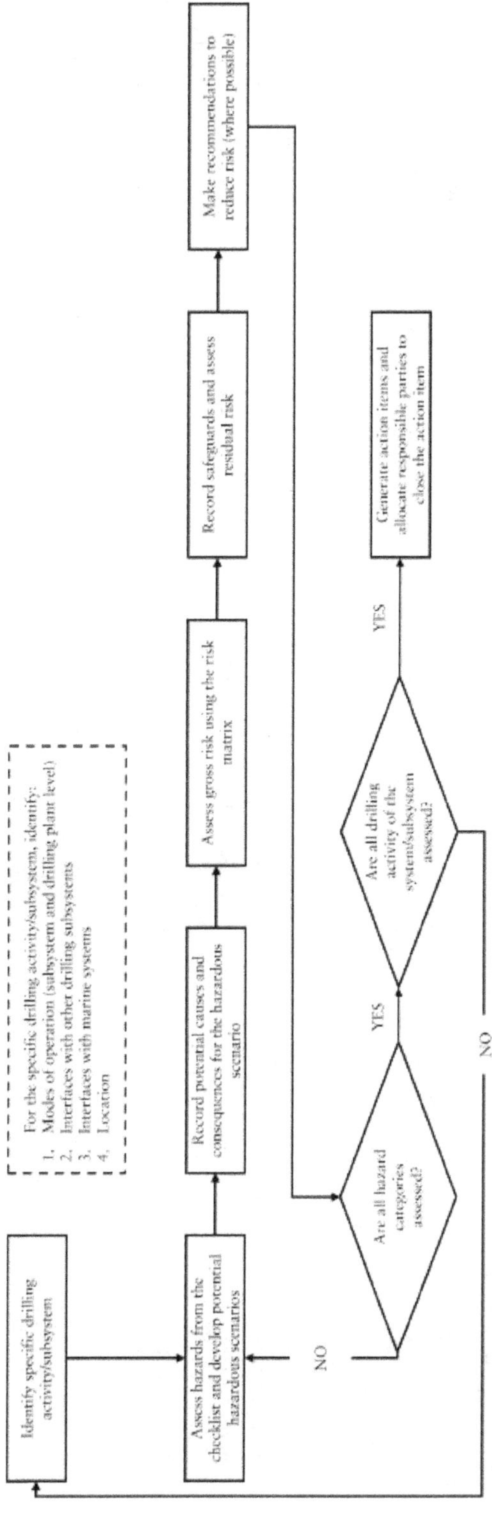

Table 7.5 Sample HAZID Table

HAZID	
Area	Mud process room and Shale Shaker Room
Major Equipment in area	Shale shaker, gumbo conveyor, APV, degasser, flow divider, mud process control room
Modes of Operation	Drilling Operations
Interfaces with other drilling subsystems	Cyber base and control system
Interfaces with marine systems	Electrical system, utility system, sea water, etc.
Reference Docs/Dwgs	GA

Hazardous Situations	*Causes*	*Consequences*	*Safeguards*	*Recommendations*	*Comments*
Release of H2S, CO2, hydrocarbons, etc., into the mud process room and shale shaker room.	Formation gases coming back.	• Potential for injury or loss of life • Fire and Explosion • Equipment failure • Of special concern is the location of the operator's control room within the mud process area.	• H2S gas detector • Zone certified equipment • Negative pressure in Shaker room • Positive pressure in operator's control room • Air change ration 12x • Crew Operational training H2S • Fire fighting/detection arrangement	Consider a fire and gas and an evacuation study to determine if there is need to relocate operator's control room or add additional safeguards.	The location of the manned operator's control room within the mud process area is of concern as there is a significant likelihood of toxic or flammable gases in area.

Table 7.5 Sample HAZID Table (continued)

HAZID	
Area	Mud process room and Shale Shaker Room
Major Equipment in area	Shale shaker, gumbo conveyor, APV, degasser, flow divider, mud process control room
Modes of Operation	Drilling Operations
Interfaces with other drilling subsystems	Cyber base and control system
Interfaces with marine systems	Electrical system, utility system, sea water, etc.
Reference Docs/Dwgs	GA

Hazardous Situations	Causes	Consequences	Safeguards	Recommendations	Comments
			• Breathing apparatus or cascade systems as applicable • H2S operational procedures		

DUAL FUEL DIESEL ENGINES (DFDE)

Purpose of Risk Study

Risk studies are required for gas fuelled engines ships for the following two cases:

1. FMEA on instrumentation and safety systems of reliquefication unit; and
2. Risk study (HAZID, FMEA, or alike) on the dual fuel diesel and single gas fuel engines, or in the dual fuel gas turbine propulsion systems.

FMEA of Instrumentation and Safety Systems

The purpose of conducting an FMEA on the instrumentation and safety systems is to demonstrate that any single failure will not lead to an undesirable event such as:

*   Lost or degraded function beyond acceptable performance criteria; and
*   Hazardous situations (such as a state that is not fail-safe, complete loss of control, unsafe shutdown of equipment, loss of propulsion, blackout, etc.).

Corrective action measures should be proposed to address situations of no-compliance with the single-failure design philosophy and mitigate the risk of hazardous events.

FMEA of Gas Fuelled Engine/Turbine

The purpose of the risk study of the engine or turbine is to identify the major hazards to people, environment, and equipment associated with the engine and verify the adequacy of the risk control measures. The risk study shall focus on identifying hazardous situations originating from:

*   Gas leakage/loss of containment.
*   Explosion/fire; and
*   Loss of propulsion.

Appropriate corrective action measures should be proposed to mitigate the risk of hazardous events that do not have adequate risk controls. The risk analysis of the engine/turbine can be accomplished via the HAZID technique, the HAZOP techniques, or the FMEA technique or a combination of techniques.

Undesired Events

Harm to:

1. Human safety.
2. The environment; and
3. Equipment.

Systems or Subsystems

FMEA of Instrumentation and Safety Systems
Failures of components within the computer-based control system for the re liquefaction unit are to be addressed in the FMEA, including the interfaces with other systems such as I/O signals. Functional failures of the equipment under control shall also be considered as to ascertain the adequacy of the actions of the safety-related functions in case of these failures.

FMEA of Gas Fuelled Engine/Turbine
The physical boundaries or the engine/turbine risk study should include the following systems:

- Dual Fuel Diesel and Single Gas Fuel Engines.
- Control and monitoring system for above; and
- Interfaces with other systems and utilities and emergency systems (e.g., gas line, gas combustion unit, reliquefication unit, etc.).

Modes of Operation

A system normally has multiple modes of operation, and each mode can present distinct failure scenarios (failure modes, hazards, consequences). The risk studies must analyse all the modes of operations to identify hazards and consequences of concern. Modes of operations to be analysed include, but are not limited to, the following:

- Fuelling.
- Underway (laden and ballasted).
- Stand-by.
- Loading/Unloading.
- Leaving berth/manoeuvring in port/coastal passage/ocean passage.
- Normal startup/shutdown; and
- Emergency shutdown.

Typical Failures

FMEA of Instrumentation and Safety Systems
The lowest level of physical component failure required to be included in the FMEA for control systems is typically at the data acquisition unit level which includes the CPU, its I/O modules, the powering of the device, and any communication between multiple devices. Physical components whose failures should be analysed in the FMEA include:

- CPU microprocessor/PLC.
- Input/output modules.
- Power supplies.
- Electrical power cables.
- Network hubs.
- Buses.
- Communication/data link cables.
- Interfaces/displays.
- Electrical power cables.
- Others, as applicable.

For functions permitting wireless data communications, consideration is to be given to the possibility of corrupted data and intermittent faults with comparatively long recovery times between interruptions. Software may reside in several different areas of a given system and at multiple levels (i.e., operating system, PLC embedded logic). Software failures are to be considered at a higher level, almost like a system failure; to address how the system can prevent and recover from, for example, malware. Classification Rules require the system developer to follow a recognised software quality assurance programme, so it is generally assumed that by the time the software is delivered and installed, it is error-free. Relays, terminal boards, indicator lights, switches, meters, and instruments do not have to be included in the FMEA unless their failure leads to a hazardous situation not discussed with a higher-level component.

FMEA of Gas Fuelled Engine/Turbine

The risk assessment workshop shall identify potential hazardous situations and verify that adequate risk control measures are provided. The list of hazards will be the guide to assist the team in identifying potential loss scenarios. The suggested list of hazards to be considered in the analysis are as follows:

- Gas leakage/loss of containment.
- Explosion/fire; and
- Loss of propulsion.

In addition, the risk assessment is to analyse issues regarding integration and interconnections between the Dual Fuel Diesel Engine or Single Fuel Gas Engine, utilities, the reliquefication unit, GCU, etc.

Timeline/Team

The risk assessment study should be performed early enough in the design stage to facilitate the implementation of recommendations that may involve a design change.

FMEA of Instrumentation and Safety Systems

The sponsor of the FMEA will be the shipyard in conjunction with the vendor of the reliquefication unit and the vendor of the control system (if different). Subject matter experts in appropriate disciplines are key to the successful identification of hazards and appropriate development of hazardous scenarios. Participants in the risk assessment workshop include subject matter experts from:

- Vendor (fuel gas supply system, dual fuel diesel engine, reliquefication unit, gas combustion unit).
- Operations.
- Shipyard.
- Control system integrator.
- Safety/HSE; and
- Team facilitator and scribe with extensive experience facilitating/recording risk assessment studies.

FMEA of Gas Fuelled Engine/Turbine

The sponsor of the engine FMEA is the engine manufacturer. Ideally, the FMEA team would include not only specialists in the engine, but in operations and in systems that the engine may interface with such as reliquefication unit and gas combustion unit.

Verification Programme

Not applicable.

Supporting Documents

The intent and scope of these requirements are the same as those for Gas Fuelled Engines. Refer to Chapter 7.

Lifecycle Management

Risk studies results are used as an aid to demonstrate the safety of the design. If any subsequent modifications are to be conducted, the risk studies are to be updated to demonstrate that any hazards derived from the modifications have been mitigated. These risk studies shall be resubmitted to Class after a modification.

Additional Notes

It is a recommended practice to use a risk matrix to qualitatively assess the consequence and the likelihood of the hazardous situations identified in the risk assessment. This risk ranking will aid in the decision as to what type of corrective actions, if any, are needed. More information on this subject is in Chapter 2. Both the FMEA for controls and safety systems and the risk study for the engine/turbine can benefit from explicitly addressing criticality via a risk matrix to allow a clearer and transparent focus on the most critical items.

GAS-FUELLED ENGINES

Purpose of Risk Study

Risk studies are required for gas fuelled engines ships for the following two cases:

- FMEA on instrumentation and safety systems for fuel gas (FG) supply system, reliquefication unit, and gas combustion units (GCU); and
- Risk study (HAZID, FMEA, or alike) on the dual fuel diesel and single gas fuel engines, or in the dual fuel gas turbine propulsion systems.

FMEA of Instrumentation and Safety Systems
The purpose of conducting an FMEA on the instrumentation and safety systems is to demonstrate that any single failure will not lead to an undesirable event such as:

- Lost or degraded function beyond acceptable performance criteria; and
- Hazardous situations (such as a state that is not fail-safe, complete loss of control, unsafe shutdown of equipment, loss of propulsion, blackout, etc.

Corrective action measures should be proposed to correct situations of non-compliance with the single-failure design philosophy and mitigate the risk of hazardous events.

FMEA of Gas Fuelled Engine/Turbine
The purpose of the risk study of the engine or turbine is to identify the major hazards to people, environment and equipment associated with the engine and verify the adequacy of the risk control measures. The risk study shall focus on identifying hazardous situations originating from:

- Gas leakage/loss of containment.
- Explosion/fire; and
- Loss of propulsion.

Appropriate corrective action measures should be proposed to mitigate the risk of hazardous events that do not have adequate risk controls. The risk analysis of the engine/turbine can be accomplished via the HAZID technique, the HAZOP techniques, or the FMEA technique or a combination of techniques.

Consequences of Concern

Harm to:

1. Human safety.
2. The environment; and
3. Equipment.

Systems or Subsystems

FMEA of Instrumentation and Safety Systems
Failures of components within the computer-based control system for:

1. Fuel gas supply system.
2. Reliquefication unit; and
3. Gas combustion units/thermal oxidisers are to be addressed in the FMEA, including the interfaces with other systems such as I/O signals.

Functional failures of the equipment under control shall also be considered as to ascertain the adequacy of the actions of the safety-related functions in case of these failures.

FMEA of Gas Fuelled Engine/Turbine
The physical boundaries or the engine/turbine risk study should include the following systems:

*   Dual Fuel Diesel and Single Gas Fuel Engines.
*   Control and monitoring system for above; and
*   Interfaces with other systems and utilities and emergency systems (e.g., gas line, gas combustion unit, reliquefication unit, etc.).

Modes of Operation

A system normally has multiple modes of operation, and each mode can present distinct failure scenarios (failure modes, hazards, consequences). The risk studies must analyse all the modes of operations to identify hazards and consequences of concern. Modes of operations to be analysed include, but are not limited to, the following:

*   Fuelling.
*   Underway (laden and ballasted).
*   Stand-by.
*   Loading/Unloading.
*   Leaving berth/manoeuvring in port/coastal passage/ocean passage.
*   Normal startup and shutdown; and
*   Emergency shutdown.

Typical Failures

FMEA of Instrumentation and Safety Systems
The lowest level of physical component failure required to be included in the FMEA for control systems is typically at the data acquisition unit level which includes the CPU, its I/O modules, the powering of the device, and any communication between multiple devices. Physical components whose failures should be analysed in the FMEA include:

*   CPU microprocessor/PLC.
*   Input/output modules.

- Power supplies.
- Electrical power cables.
- Network hubs.
- Buses.
- Communication/data link cables.
- Interfaces/displays.
- Electrical power cables; and
- Others, as applicable.

For functions permitting wireless data communications, consideration is to be given to the possibility of corrupted data and intermittent faults with comparatively long recovery times between interruptions. Software may reside in several different areas of a given system and at multiple levels (i.e., operating system, PLC embedded logic). Software failures are to be considered at a higher level, almost like a system failure; to address how the system can prevent and recover from, for example, malware. Classification Rules require the system developer to follow a recognised software quality assurance programme, so it is generally assumed that by the time the software is delivered and installed, it is error-free. Relays, terminal boards, indicator lights, switches, meters, and instruments do not have to be included in the FMEA unless their failure leads to a hazardous situation not discussed with a higher-level component. Functional failures of the equipment under control shall also be considered as to ascertain the adequacy of the actions of the safety-related functions in case of these failures.

FMEA of Gas Fuelled Engine/Turbine
The risk assessment workshop shall identify potential hazardous situations and verify that adequate risk control measures are provided. The list of hazards will be the guide to assist the team in identifying potential loss scenarios. The suggested list of hazards to be considered in the analysis are as follows:

- Gas leakage/loss of containment.
- Explosion/fire; and
- Loss of propulsion.

In addition, the risk assessment is to analyse:

- Issues regarding integration and interconnections between the Dual Fuel Diesel Engine or Single Fuel Gas Engine, utilities, the reliquefication unit, GCU, etc.

Timeline/Team

The risk assessment study should be performed early enough in the design stage to facilitate the implementation of recommendations that may involve a design change.

FMEA of Instrumentation and Safety Systems
The sponsor of the FMEA will be the shipyard in conjunction with the vendor of the particular equipment (Fuel Gas Supply system, reliquefication unit, gas combustion units) and the vendor of the control system (if different). Subject matter experts in appropriate disciplines are key to the successful identification of hazards and appropriate development of hazardous scenarios.

Participants in the risk assessment workshop include subject matter experts from:

- Vendor (fuel gas supply system, dual fuel diesel engine, reliquefication unit, gas combustion unit).
- Operations.
- Shipyard.
- Control system integrator.
- Safety/HSE; and
- Team facilitator and scribe with extensive experience facilitating/recording risk assessment studies.

FMEA of Gas Fuelled Engine/Turbine
The sponsor of the engine FMEA is the engine manufacturer. Ideally, the FMEA team would include not only specialists in the engine, but in operations and in systems that the engine may interface with such as reliquefication unit and gas combustion unit.

Verification Programme

Not applicable.

Supporting Documents

The risk assessment report must be submitted to Class for review. In order to conduct a proper review, the Class Plan reviewer needs the following information:

- Completed risk assessment report including the following:
  - Description of objective and scope.
  - Description of hazard identification technique used.
  - List of team members who participated in the workshop.
  - Description of each action item and responsible party to close the item, as identified by the team during the workshop.
- Completed risk assessment workshop record describing the potential accident scenarios discussed by the team as well as corrective actions.
- Detailed description, narrative and drawings of the system functional description and control.
- Layout and general arrangement drawings.
- System design specifications.
- Cause and effect chart showing the control logic.
- Description of the physical and operational scope and boundaries for the risk study; and
- Operations, maintenance, inspection, and testing manuals.

The operations, maintenance, inspection, testing, and emergency manuals are needed for Class to confirm compliance with the design and manufacturer's requirements and that the findings of the FMEA have been incorporated into the equipment operating phases.

Lifecycle Management

Risk studies results are used as an aid to demonstrate the safety of the design. If any subsequent modifications are to be conducted, the risk studies are to be updated to demonstrate that any hazards derived from the modifications have been mitigated. These risk studies shall be resubmitted to Class after a modification.

Additional Notes

It is a recommended practice to use a risk matrix to qualitatively assess the consequence and the likelihood of the hazardous situations identified in the risk assessment. This risk ranking will aid in the decision as to what type of corrective actions, if any, are needed. More information on this subject is in Chapter 2. Both the FMEA for controls and safety systems and the risk study for the engine/turbine can benefit from explicitly addressing criticality via a risk matrix to allow a clearer and transparent focus on the most critical items.

MOTION COMPENSATION AND ROPE TENSIONING SYSTEMS FOR CRANES

Purpose of FMEA

Demonstrate that a single failure of the motion compensation systems and rope tensioning systems for cranes will not cause loss of control of load.

Undesired Events

Loss of control of load, structural damage, or other hazardous situations such as fire and explosion, pollution, or injury.

Systems or Subsystems

Motion compensation systems and rope tensioning systems for cranes.

Modes of Operation

The FMEA is to consider failure during all foreseeable crane modes of operation.

Typical Failures

Typical failures that need to be analysed in the FMEA include, but are not limited to:

- Electric failures.
- Hydraulic failures.
- Mechanical failures.
- Failure of crane power supply from vessel.
- Failures of lifting active equipment components.
- Failures of passive components such as pipes, cables, block valves, etc.
- Failures of control systems at I/O level (e.g., sensors, power supply, PLC, cables, etc.).
- Failures at interfaces or interconnections between systems.
- Failures caused by credible external events; and
- Others, as applicable.

Particular attention should be paid in the FMEA to the interfaces between systems.

Timeline/Team

- The optimal time for performing the FMEA early design is at the vendor stage when there is still time to include FMEA recommendations in the final design and integration process.

- Ideally, subject matter experts in operations, design and relevant vendors should participate in the FMEA; and
- The FMEA report shall be submitted to Class with approval drawings.

Verification Programme

There are no specific Classification requirements for testing to validate the FMEA conclusion. However, it is at the discretion of the Class Surveyor or Class Plan reviewer to recommend specific testing of FMEA failures for which there is a higher degree of uncertainty.

Supporting Documents

The FMEA must be submitted to Class for review. In order to conduct a proper review of the FMEA, the Class Plan reviewer needs the following information included in the FMEA report:

- General description of the scope and purpose of the FMEA.
- Dates when the FMEA was conducted.
- FMEA participants.
- Summary of any risk reducing actions performed or pending as a result of recommendations of the FMEA.
- System drawings, including layout/general arrangement, schematics.
- Functional block diagram showing systems interactions, parallel paths for redundant systems, in series for single systems.
- Description of the system boundaries, both physical and operational.
- System design specifications.
- FMEA worksheets, including:
  - Modes of operations.
  - All significant failure modes associated with each operational mode.
  - Cause associated with each failure mode.
  - Common cause failures.
  - Method for detecting failure has occurred.
  - Effect of the failure upon the system.
  - Global effect of failure on other systems, the asset and HSE; and
  - Controls to prevent and/or mitigate failure.

The operations, maintenance, inspection, testing, and emergency manuals are needed for Class to confirm compliance with the design and the manufacturer's requirements and that the FMEA findings have been incorporated into the equipment operating phases.

Lifecycle Management

If any subsequent modifications to the cranes are conducted, risk studies are to be updated to demonstrate that the system is still in compliance with the Class design philosophy. Types of changes requiring submission of the revised FMEA for Class review include:

- Operating criteria changes.
- System configuration/boundary changes.
- Equipment changes.
- New application; and
- Revalidation of previous FMEA from vendor.

Additional Notes

Nil.

Annex

Sample FMEA and FMECA Worksheets

SAMPLE FMEA/FMECA WORKSHEETS

This appendix provides sample worksheets to illustrate the characteristics of an FMEA or a FMECA requested to demonstrate compliance with Classification. Only one or two representative pages of an FMEA or a FMECA are provided for illustration. The examples shall by no means be considered a full worksheet for the system nor shall they be construed to be the only suitable approach and format. There is no significant variation among the characteristics expected for each of the FMEA or FMECA required for Classification. However, certain requirements are more defined than others, and this annex intends to show those variations. The sample worksheet shown is for FMEA for the BOP Control System which would satisfy ISQM as well as CDS FMEA/FMECA requirements.

FMECA WORKSHEET EXAMPLE (FOR ISQM AND FOR CDS)

The example shown in Table A.2 illustrates select sections of the FMECA worksheet for the BOP control system on a drilling unit. This FMECA would satisfy both the requirements of the ISQM Guide and the CDS Guide. The ISQM FMECA requirements are well defined, as indicated in Table A.1, and include specific suggestions on how to document the FMECA. Table A.1 below demonstrates the typical BOP control systems, their expected Integrity Levels, and selected typical functions for each. It is recommended that the functions be listed in a separate table and that each function be given a unique identifier. Note that the determination of Integrity Levels or the criticality is not a requirement of the CDS Guide, only of ISQM.

Table A.1 Example of BOP Control Functional Items

BOP Control System	Integrity Level	Functions
Power/Communication Distribution Cabinet Y	IL3	F1. Well shut-in
		F2. Subsea communication
		F3. Closure of Riser Gas Handling Annular
		F4. Casing emergencies disconnect (EDS)
		F5. Drill pipe emergency disconnect (EDS)
		F6. Controlled release of riser tension via Riser Recoil system during emergency disconnect t
Power/Communication Distribution Cabinet Z	IL3	Each subsystem has one or more functions that need to be determined. In this example, functions are illustrated only for the Power/ Communication Distribution Cabinet Y above
Driller's Control Panel	IL3	...
Bridge Control Panel	IL3	...
Diverter Control Unit	IL3	...
Subsea Electronics Module	IL3	...
Automatic Mode Function	IL3	...
Emergency Disconnect	IL3	...
Riser Message String Interface	IL2	...
Network Interfaces	IL3	...
Event Logger, Remote Logging server, BOP Test system, Acoustic Function	IL1	N/A. Outside scope of FMEA.

Table A.2 FMECA Worksheet Example (Select Sections of a FMECA for BOP Control System)

Operational Mode												
Unit												
Description of Unit		Description of Failure			Effects of Failure		Safeguards		Sever-ity	Likeli-hood	Risk	Corrective Action	Responsi-ble
Function	ID #	Failure Mode	Failure Causes	Detec-tion of Failure	Local	Global	Prevention of Failure	Mitigation of Effect	L, M, or H	L, M, or H	L, M, or H		
...	...	...	...	...	...	...	...	...	...	...	...	...	...
F5: Drill pipe emer-gency discon-nect (EDS)	PDCX5.1	Lack of function-ality: EDS does not occur on demand upon a well control event	Sensor fail-ure, code error, commu-nication failure	Visual	None	Potential escalation of well control event. Major safety, en-vironmen-tal and business	Software review, software test, func-tional test, preventive mainte-nance	Shear rams				Safe-guards consid-ered ad-equate	

Table A.2 FMECA Worksheet Example (Select Sections of a FMECA for BOP Control System)

| Operational Mode | | | | | | | | | | | | | |
| Unit | | | | | | | | | | | | | |
Description of Unit		Description of Failure			Effects of Failure		Safeguards		Sever-ity	Likeli-hood	Risk	Corrective Action	Responsi-ble
Function	ID #	Failure Mode	Failure Causes	Detec-tion of Failure	Local	Global	Prevention of Failure	Mitigation of Effect	L, M, or H	L, M, or H	L, M, or H		
F5: Drill pipe emergency disconnect (EDS)	PDCX5.2	False action: EDS initiates spuriously	Sensor failure, code error, communication failure	Visual	Interruption of drilling	Potential for safety, environmental and business impacts	Software review, software test, functional test, preventive maintenance	Manual Override				Safeguards considered adequate	
F5: Drill pipe emergency disconnect (EDS)	PDCX5.3	Improper functionality: Erroneous EDS sequence	Sensor failure, code error, communication failure		None	Potential for safety, environmental and business impacts	Software review, software test, functional test, preventive maintenance					Safeguards considered adequate	
F5: Drill pipe emergency disconnect (EDS)	PDCX5.4	Timing: Incorrect timing of valve commands upon ESD demand	Code error, communication failure	Visual indication at event logger	None	Potential for safety, environmental and business impacts	Software review, software test, functional test	None	High	Low	Medium	Safeguards considered adequate	

Table A.2 FMECA Worksheet Example (Select Sections of a FMECA for BOP Control System)

Operational Mode													
Unit													
Description of Unit		Description of Failure			Effects of Failure		Safeguards		Sever-ity	Likeli-hood	Risk	Corrective Action	Responsi-ble
Function	ID #	Failure Mode	Failure Causes	Detec-tion of Failure	Local	Global	Prevention of Failure	Mitigation of Effect	L, M, or H	L, M, or H	L, M, or H		
F5: Drill pipe emergency disconnect (EDS)	PDCX5.5	Sequence: Wrong valving sequence (upon ESD demand)	Code error, communication failure	Visual indication at event logger	None	Potential for safety, environmental and business impacts	Software review, software test, functional test	None	High	Low	Medium	Safeguards Considered Adequate	
F5: Drill pipe emergency disconnect (EDS)	PDCX5.6	Memory Management :EDS errors	Memory errors	Visual, audible alarm	Potential for crash of PLC-X	Loss of PLC redundancy	Software review, software test, functional test	Redundant PLC-Z	Low	Medium	Low	Safeguards Considered Adequate	
F6: Controlled release of riser tension via Riser Recoil System during EDS	PDCX6.1	Lack of functionality: Controlled release of riser tension did not occur during an emergency disconnect	Code error, interface error, communication error	Visual at event logger and HMI lack of output from PLC	None	Potential safety and business impacts	Software review, software test, functional test, signals	None	High	Low	Medium	Safeguards Considered Adequate	

Table A.2 FMECA Worksheet Example (Select Sections of a FMECA for BOP Control System)

Operational Mode

Unit

Description of Unit		Description of Failure			Effects of Failure		Safeguards		Severity	Likelihood	Risk	Corrective Action	Responsible
Function	ID #	Failure Mode	Failure Causes	Detection of Failure	Local	Global	Prevention of Failure	Mitigation of Effect	L, M, or H	L, M, or H	L, M, or H		
F6: Controlled release of Riser Tension via Riser Recoil System during EDS	PDCX6.2	False action: spurious controlled release of riser tension	Code error, interface error, communication error	Visual at event logger and HMI	None	Potential safety and business impacts	Software review, software test, functional test, signals	None	High	Low	Medium	Safeguards considered adequate	
...	...	...	...	...	...	...	...	...	...	...	...	...	...

About the Nørdik Press

The Nørdik Press is an independent publisher of maritime and ship related books. Drawing on the heritage of the historic maritime city of Southampton, Nørdik Press is proud to work with new and established authors from across the industry, wherever in the world they may be.

About the Author

Alexander Arnfinn Olsen is an established author and training professional. He has worked in the maritime training profession for over 20 years and acquired a wealth of knowledge and experience, which is shared through his many publications.

SELECTED BOOKS AND TITLES BY THIS AUTHOR

Books published by Routledge

- Introduction to Ship Engine Room Systems
- Maritime Accident and Incident Investigation
- Maritime Cargo Operations
- Merchant Ship Types
- Core Principles of Maritime Navigation
- Firefighting and Fire Safety Systems on Ships
- Introduction to Container Ship Operations and Onboard Safety

Books published by Springer

- Ship Operations in Extreme Low Temperature Environments
- Introduction to Digital Navigation
- Introduction to Celestial Navigation

Books published by the Magellan Maritime Press

- Ship Sanitation and Hygiene